ABC Family t

ABC Family to Freeform TV

*Essays on
the Millennial-Focused
Network and Its Programs*

Edited by EMILY L. NEWMAN
and EMILY WITSELL

McFarland & Company, Inc., Publishers
Jefferson, North Carolina

ISBN (print) 978-1-4766-6735-5
ISBN (ebook) 978-1-4766-3216-2

LIBRARY OF CONGRESS CATALOGUING-IN-PUBLICATION DATA

BRITISH LIBRARY CATALOGUING DATA ARE AVAILABLE

Front cover image of people watching TV © 2018 4x6/iStock

Printed in the United States of America

McFarland & Company, Inc., Publishers
Box 611, Jefferson, North Carolina 28640
www.mcfarlandpub.com

To Charlie, Dean, and Fred,
for being our tireless companions,
and never leaving our sides.

A special thank you to all of our contributors.

ELN and EW

To the women who both inspire me and put up
with my constant yammering on about television:
Laurie and Allison Newman. For my Mema, June Wright,
who will always be my inspiration.
ELN

To my mother and father, Mary and Ed Witsell,
who have always supported my strange and wonderful interests,
and who gave me a love of books that led me to create one myself.
EW

Table of Contents

A New Kind of Family

Identity Issues and "Becomer" TV

Introduction

EMILY L. NEWMAN *and* EMILY WITSELL

On January 12, 2016, ABC Family officially changed its name to Freeform TV, launching the latest in a series of network-rebranding name changes for the channel that began its life as the Christian Broadcasting Network in 1977. Network executives wanted to move away from ABC Family and the "family" brand that they had been contractually obligated to use since the network was the Christian Broadcast Network Family Channel in 1988, even though it is now owned by the Walt Disney Company. In the new, crowded television landscape, children and their families would no longer be the network's main audience. Now, Freeform would target young men and women between the ages of 18 and 34, a group they deemed "Becomers."[1] According to network president Tom Ascheim, this demographic is at the beginning of their adult lives, and as such, Becomers are often on the cusp of experiencing big changes such as buying a new car, going off to and graduating from college, starting their professional careers, and other markers of adulthood. In order to capture the eyes (and advertising dollars) of this growing market, Freeform's programming now focuses on these themes; the shows currently running on the network spotlight older young adults (often in their late teens and twenties) and explore these transitional experiences across genres. As the network grew up, the shows followed suit.

The year 2016 would not be the network's first attempt at rebranding. Long before Becomers were the target audience and Freeform was the name of the network, the Christian Broadcasting Network (CBN) company and Pat Robertson founded the CBN Satellite Service cable network in 1977. The network was anchored by its signature program, *The 700 Club*,[2] which gives daily news updates with a conservative Christian bent. It was first broadcast in 1966, and is actually still aired consistently on Freeform due to contractual obligations.[3] Big changes came to the network in 1987, as Tim Robertson (Pat Robertson's son) took over and renamed the channel the CBN Family

Channel, then the Family Channel, which better communicated that parents and children could enjoy the programming together. Ten years later, another big change came about as the channel was sold as part of a larger deal to Fox Family Worldwide and the name was subsequently changed to Fox Family, hoping to compete with Nickelodeon and the Disney Channel. The network scheduled syndicated children's programming during the day and family programs in the evening, eventually rolling out their own original programming aimed at young audiences. Yet, success did not follow: past viewers were disappointed with the changes and stopped watching; new viewers never seemed to materialize. The Walt Disney Company purchased the channel in 2001 and changed its name to ABC Family, with the intention of using the channel as a home for reruns of ABC's programming. Interestingly, the Disney company initially explored naming the network XYZ, to indicate the audience (generations X, Y, and Z) that it hoped to attract, but these efforts were stymied by the contractual obligations that required that "family" remain in the network's moniker.[4] Thus, the early years of this rebranded network saw the network re-airing many of ABC's programs to accompany its minimal original programming, all the while carrying the "family" name that it had hoped to deemphasize. By syndicating many of the long-running WB dramas, coupled with older, popular ABC sitcoms, ABC Family was slowly able to shift its demographic from children and parents to young adults.[5]

By 2007, ABC Family was secure in the knowledge that its prime audience was a generation that was beginning to garner attention in the media: millennials. To introduce this idea to the public, the network sponsored the publication of a branded content item in the publication *Advertising Age*, "Getting to Know the Millennials."[6] This advertorial aimed to help advertising agencies see the value in marketing to this age demographic, which also happened to be ABC Family's core audience, while also giving the reader clues as to the network's new values. The network claimed that millennials enjoyed family time, but also wanted relevant media representations: their audience was looking for what the network called "a new kind of family," one that went beyond traditional or conservative definitions.[7]

The channel struggled to find its footing with original programming in the early years, yet when they defined their new audience and goals, success quickly followed. With *The Secret Life of the American Teenager* (2008–2013) and *Pretty Little Liars* (2010–2017), ABC Family seemed to have finally found its groove. In 2010, ABC Family saw an average of 580,000 viewers from the age of 12 to 34, up 49 percent from 2005. Just like that, ABC Family was one of cable's top ten networks for adults aged 18 to 49.[8] As their programming evolved and the network began to evaluate its real successes, it became clear that the network's strengths lay in moving away from their original children's and family-friendly programming to that which targeted an older crowd. But the audience ABC Family was attracting was not just teenagers; it included older young adults, up to the age of 34. Significantly, this is often the target audience for advertisers. It was the time for another rebranding, and this took the form of the name change to Freeform, which Ascheim explained, "spoke to core parts of our essence. It seemed to evoke an emotional response in us and our creative selves and made us feel like we would be able to convey the spirit of Freeform to the audience, and they'd want to play along with us."[9] In 2016, Freeform was born.[10]

Despite its name change and the introduction of the "Becomer" audience, in its early years, Freeform has continued to focus on the themes that made ABC Family successful: programming featuring young people from a variety of backgrounds, facing life changes in settings both realistic and fantastical. Perhaps ABC Family/Freeform's most consistently successful programming has been the primetime soap opera, which has been a staple of the network for years. In the past few years, these shows have dealt with increasingly adult themes, including murder, scandal, sex, and trauma. When the network debuted their new name and brand Freeform in 2016, one of the first shows to air new episodes was their flagship drama *Pretty Little Liars*. Yet, things were not normal in Rosewood, the fictional town where the series takes place: the show came back from hiatus with a major time jump. Five years had lapsed and the leading characters were no longer in high school; they were adults, finished with college and well on their way to adulthood. Perfectly symbolizing the network's new Becomer themes, *Pretty Little Liars* demonstrated the target audience's increasing interest in adulthood, as well as the sexier and edgier themes that Freeform wanted to explore with an older, less family-oriented brand.

Yet, all good things must come to an end, and it was announced that Freeform's well-performing *Pretty Little Liars* would end in 2017, leaving the network with a significant gap that would have to be filled with new programming. *Pretty Little Liars* had bridged the transition from ABC Family to Freeform, consistently ranking as the channel's number one show.[11] Further, it had established a model of success that notably included social media (becoming the most tweeted-about show on television ever), which the network would try to replicate with their new shows.[12] While lacking the fantastical thriller elements of *Pretty Little Liars*, family dramas *The Fosters* (2013–2018) and *Switched at Birth* (2011–2017) premiered shortly after *Pretty Little Liars* and gained loyal audiences that strengthened the network's reputation with young adults. In the summer of 2016, Freeform made two attempts to recapture the suspenseful mystery/thriller aspect of *Pretty Little Liars* with *Guilt* and *Dead of Summer*, while also premiering the drama *Recovery Road*. Unfortunately, all three shows stalled and suffered from poor ratings, resulting in cancellations after only one season.

In many ways, then, the latter half of 2017 and 2018 represent Freeform's real beginning, as it is the first time that they have been forced to replace the programs that had been successful for many years. In a smart attempt to find ratings gold, Freeform hired I. Marlene King, creator of *Pretty Little Liars*, to create another show, *Famous in Love* (2017–present), which attempts to replicate the heightened drama and intense friendships of her early show. The show, however, has struggled to find its footing—despite being scheduled following *Pretty Little Liars*' last season and debuting its entire season on Hulu on the night of its network premiere. Despite *Famous in Love*'s unsteady ratings, Freeform greenlit a second season. *Famous in Love*, *Guilt*, *Dead of Summer*, and *Recovery Road* were created after *Pretty Little Liars*, and each attempted to incorporate similar drama and intrigue, as well as utilize marketing across social media platforms. So far, none of the shows have quite captured the magic that *Pretty Little Liars* did, demonstrating that securing the Becomer audience requires more than just drama and savvy marketing.

Beyond soap operas, Freeform also explores dramatic programming that leans towards science fiction. *Stitchers* (2015–2017) sees its protagonist, Kirsten Clark (Emma Ishta), explore the memories of the recently deceased, though the real focus is on her finding friends and allies for the first time in her life. Similarly, *Beyond* (2017–present) features a young man who awakens from a 12-year coma to find himself imbued with special powers but in the midst of a complicated conspiracy. Both shows have a few mystical or unusual elements, but often rely on the development of friendship and familial bonds in the context of plots concerned with basic battles between good and evil. These science fiction–lite shows allow for a supernatural threat while still feeling very human and relatable to their audience.

Science fiction is explored more thoroughly in *Shadowhunters* (2016–present) and the upcoming *Siren* (scheduled to debut in 2018). *Shadowhunters*, adapted from a young adult book series by Cassandra Clare (real name Judith Lewis née Rumelt), explores a secret, hidden underworld of demons, fairies, hunters, and battles for power and immortality. Taking the fantasy underwater, *Siren* will focus on a potentially murderous mermaid, and the sailors and city who struggle to confront her mythology and existence. These shows in particular follow in the footsteps of *Kyle XY* (2006–2009) and *The Nine Lives of Chloe King* (2011), which had varied success on the network, but laid the groundwork for the network consistently airing at least one fantasy program at all times.

Superhero narratives and shows based on comic books have figured somewhat frequently on the network, though ABC Family never quite seemed to find much success; one effort, *The Middleman* (2008), was cancelled after only one season. Though various incarnations and new seasons of the popular *Power Rangers* franchise aired in the early 2000s, these shows were clearly targeted to a younger audience. Freeform looks to change that by cashing in the current multi-network trend toward creating series based on existing comic book franchises with two upcoming shows that are part of the Marvel universe. *Cloak and Dagger* (scheduled for 2018) features a young man and woman from different backgrounds, but who are connected to each other as both harbor powerful secrets. Also developed for a 2018 debut, Marvel's *New Warriors* focuses on a group of six people with unique abilities who live and work together. The show has garnered a fair amount of press attention, mainly because it will include the popular Squirrel Girl, a character often seen to embody female empowerment in the Marvel comic books. In fact, that was one of the main motivations behind Freeform's attempt to get involved with Marvel, and particularly feature this program, as Karey Burke (Executive Vice President of Programming and Development at the network) elaborates: "We are a network for young adults, but we're proud we're No. 1 with young women, and we want to find characters who speak to them. There's a reason we chased Squirrel Girl: she embodies all of that."[13]

In fact, the network has effectively and consistently pursued programming that appeals to young women, often by featuring unique, sophisticated, and powerful young women, such as *Make It or Break It* (2009–2012) and *Bunheads* (2012–2013), to name just a few. Further, it seems natural that Squirrel Girl and *New Warriors* will follow the release of *The Bold Type* (2017–present), which similarly works to embolden girls and young women, without the fantastical settings and elements. The show is executive produced by Joanna Coles, former editor-in-chief of *Cosmopolitan* magazine, and details the stories of three women at the beginning of their careers as they try to succeed in life and love. Promoting strong female friendships and supportive bosses, the show feels like an anomaly when compared to much of what appears on television today, which rarely features as many strong female characters as this, and if so they are often pitted against each other, as seen in *Game of Thrones* (HBO, 2011–present), *Crazy Ex-Girlfriend* (The WB, 2015–present), *Reign* (The WB, 2013–2017), *Younger* (TV Land, 2015–present), and *Girlfriends' Guide to Divorce* (Bravo, 2014–present), among other examples. In *The Bold Type*, the trio of friends are constantly helping each other, providing encouragement, and working together so that each can achieve their goals. Beyond that, they have an intense but generous leader, as creator and showrunner Sarah Watson explains: "We always see on TV the women who are tough and not rooting for other women to succeed, and that's not the kind of mentors I've had in my life. I've had incredible female bosses, and I wanted to show someone one wants to bring up the next generation of strong women writers."[14] *The Bold Type* incorporates robust friendship bonds like *Pretty Little Liars*, yet is set in a creative and nurturing environment. Of course, the women deal with plotlines that force them out of their comfort zones and see them struggle to find their voices, yet with the help of each other and their benevolent (but tough) boss, the young women are able to succeed.

Independent women also feature prominently in the current and upcoming sitcoms on Freeform, though, expectedly, they often find themselves in zany and wacky scenarios. *Young and Hungry* (2014–present) is loosely based on real life food blogger Gabi Moskowitz. Aspiring chef Gabi Diamond becomes a personal chef to a wealthy, young entrepreneur, with whom she predictably falls in love. Two upcoming shows for 2018 also seem to focus on empowering young women: *Grown-ish* and *Alone Together*. A spin-off of popular ABC sitcom *Black-ish* (2014–present), *College-ish* follows the eldest Johnson daughter as she begins college. Fronted by Yara Shahidi, a vocal feminist, the show promises to build on *Black-ish*'s strong, politically oriented point-of-view that often encourages its audience to consider themes of social justice. Created by Lonely Island (a comedic troupe composed of Andy Samberg, Akiva Schaffer, and Jorma Taccone), *Alone Together* features two platonic best friends who support each other as they try to survive Los Angeles.[15] Each of the above-mentioned shows epitomize the idea of the Becomer—focusing on transitions and major life developments.

While ABC Family/Freeform may not be known for their reality programming, they consistently try their hand at it, and in the past years their choices have increasingly worked to support the Becomer viewer. While *Monica the Medium* (2015–2016) focused on a college-aged woman coming into her career as a psychic, this is just one of many reality programs that deal with young people figuring out their professional goals and careers, like *Job or No Job* (2015), *Next Step Realty: NYC* (2015), and *Startup U* (2015), among others. Both *Cheer Squad* (2016) and *The Letter* (2016) showed women (and a few men) who were working to become better in their life, either in cheerleading or in love and friendship, respectively. Additionally, the Walt Disney company sought to capitalize on their popular ABC *Bachelor* franchise with two new programs: *Ben and Lauren: Happily Ever After?* (2016, following a former *Bachelor* contestant and his fiancée) and *The Twins: Happily Ever After?* (2017, featuring two former contestants from *The Bachelor* season 20). While these shows allow for Freeform to expand on their juggernaut franchise, the network also chose to create shows focused on people's lives in transition, as the protagonists are all trying to come to terms with a major life change. Ben, Lauren, and the twins are all, in fact, Becomers.

Now, as 2018 opens, Freeform is at a moment of transition. The stalwarts of *Pretty Little Liars* and *Switched at Birth* have ended, leaving the channel without the clear anchors they have had for years.[16] As the network continues to look for their next big hit, they will undoubtedly rely on what has made them so successful in the past: innovative marketing strategies, engagement with social media, and all the drama and intrigue young adults can handle.

Literature Review

Despite its short history, ABC Family/Freeform has attracted attention from scholars from a broad range of disciplines. This section describes several important contributions to the scholarship surrounding the network's evolution and branding, the television and media environment in which ABC Family/Freeform was founded and has come to thrive, and scholarly literature focused on individual shows and themes.

Though Freeform now prefers the term "Becomers," the network's desired audience has been the millennial generation since its ABC Family days. In her book *Millennial Fandom: Television Audiences in the Transmedia Age*, Louisa Ellen Stein describes the rise of ABC Family and its early focus on millennials, detailing how ABC Family latched onto the concept of the socially conscious, diverse millennial as its target audience. Stein also details ABC Family's early adoption of technology as an opportunity to build brand loyalty, but concludes that they did so on their own terms: while the network welcomed fan communities and user-generated content, they limited submissions to content that dealt with friends and family, considering these topics to be the network's purview.[17] Stein explores how ABC Family co-opted existing fannish behaviors for the network's benefit so that viewers would be "engaged enough to strengthen the ABC Family brand but not disrupt it."[18]

Caryn Murphy also considers ABC Family's strategy of marketing to millennials in her essay "Secrets and Lies: Gender and Generation in the ABC Family Brand." After detailing the complex history of ABC Family from its inception as the Christian Broadcasting Network to 2014, Murphy describes how ABC Family has focused much of its attention on millennial women through creating original programming with young female leads, building on techniques employed by The WB network, from which ABC Family purchased much of its second-run programming in its early years.[19] She connects the buying power of millennial women to ABC Family's pursuit of advertising dollars via female-oriented programming, but also considers the progressive nature of many of the network's shows in terms of the inclusion of representations of diversity through race, ethnicity, disability, sexuality, and intellectual abilities.[20]

Freeform's decision to focus specifically on the Becomer generation can be seen as a part of a larger trend: cable television's movement toward narrowcasting, or the desire of television networks and advertisers to target a small, clearly defined audience. As early as 1988, the *Journal of Advertising* was writing about the need for small cable networks, which could not compete with the sheer reach and wealth of the then–Big Three networks (ABC, CBS, and NBC), to create a niche space for a specialized viewer who would develop loyalty to the network. Finding a niche market that would reach a sizeable and growing audience was, and remains, essential to the survival of cable networks, which rely on advertising dollars to remain afloat.[21]

As cable television reached more households in the early 1990s, scholars explored the effects of narrowcasting on various demographics. Heather Hundley studied the Lifetime television network, one of the earliest networks to successfully employ narrowcasting, and coined a new term for the phenomenon in "The Evolution of Gendercasting: The Lifetime Television Network— 'Television for Women.'" In her exploration of the evolution of Lifetime, Hundley concludes that narrowcasting benefits both the audience and the advertisers: audiences find that their interests are being represented on television, while advertisers can tailor their advertisements to speak directly to a specific market, rather than trying to appeal to a broad swath of the population.[22]

In 2006, Amanda Lotz examined the rise of television networks dedicated to women in the 1990s and early 2000s, questioning whether these new spaces provided room for alternative interpretations of women's roles or rather reinforced existing negative stereotypes.[23] Focusing specifically on the explosion of dramatic series helmed by at least one female character, Lotz claims that though the expanded landscape of women's entertainment allows for a plurality of stories that had not been present in the era of broadcast television, these stories are not by their very existence feminist.[24] Lotz concludes that though the rise of networks and programming designed to appeal to women was certainly a business decision aimed at earning more advertising dollars, "the cause does not render the effect meaningless for critical scholars studying hegemonic struggles in popular culture,"[25] and these changes are certainly still relevant for scholars examining the decisions made by ABC Family and Freeform, networks who hold the stated goal of appealing to Becomer women.

Lotz continued her contributions to the field of television studies in 2014 with the second edition of *The Television Will Be Revolutionized*, a study of how technology is in the process of disrupting television viewing and redefining the industry. Declaring the 2000s to be the beginnings of the "post-network era," Lotz recounts that DVDs, DVRs, streaming, and other innovations have allowed audiences to consume television outside of programming schedules, but notes that traditional viewing still accounted for the majority of television consumption as of 2014.[26] At the same time, these new technologies have led to ever-increasing ways of viewing shows and delivering ad content, with changes coming at a rapid pace. Therefore, networks and advertisers must adapt quickly, leading Lotz to wonder if the post-network era is moving from narrowcasting to the even narrower concept of person-casting, defined as "what is viewed, when, how, and even in how viewers pay for it."[27] These ideas play out in the marketing and branding of many of ABC Family's shows, particularly *Pretty Little Liars*.

Individual shows created by ABC Family and Freeform have garnered scholarly attention in their own right. Many of these studies focus on the areas that the network has described as their key priorities: diverse depictions of young people and modern marketing techniques that reach the network's young audience. What follows is a representative, though not comprehensive, sample of the scholarship currently being published on ABC Family/Freeform programming.

Melanie E.S. Kohnen writes about *The Fosters'* multiracial cast, asking whether the show rises to the bar of "meaningful cultural diversity" and whether ABC Family's use of diversity as a branding strategy leads to simplified representations of minority characters. She ultimately concludes that far from being tokenized, the show's characters are fully formed and complex. Thus, ABC Family offers viewers a space for exploration of identity through the show.[28]

A number of scholars have been drawn to ABC Family/Freeform's portrayal of gay and lesbian characters, a nod to the network's consistent commitment to depicting a range of sexualities across their programming. For example, in "The Best Lesbian Show Ever! The Contemporary Evolution of Teen Coming-Out Narratives," Jennifer Mitchell identifies *Pretty Little Liars* as a foundational show for later representations of lesbian teenagers on television thanks to its strong coming-out narrative and wealth of lesbian characters.[29] Wendy Peters looks at *Pretty Little Liars* and *The Secret Life of the American Teenager*'s out-of-the-closet gay and lesbian characters in "Bullies and Blackmail: Finding Homophobia in the Closet on Teen TV." Peters notes that the 2010–2011 television season featured more out gay characters across all channels than had been featured in the entire preceding decade. She concludes that the ABC Family shows' emphasis on the ease of coming out in high school presents being out as a positive alternative to hiding one's sexuality, ultimately encouraging young viewers to embrace their identities.[30] In 2015, Shane Brown reflected on how one of ABC Family's earliest original series, *Kyle XY*, explored gender roles and norms, determining that the show tested the waters in presenting more progressive gender roles but ultimately did not fully disrupt gender stereotypes for its characters.[31]

In the midst of the celebration of diverse representations on the network, Seon-Kyoung An et al. performed an experiment to determine whether seeing disability on television actually changes viewers' attitudes in "Prominent Messages in Television Drama *Switched at Birth* Promote Attitude Change

Toward Deafness." The researchers found that watching as little as a single episode of *Switched at Birth* could improve a number of views on deafness, though improvement was still needed in certain areas.[32] In a similar study, Traci K. Gillig and Sheila T. Murphy had adolescents watch clips of *The Fosters* that showed the evolution of the relationship between two gay teen characters. Their analysis showed that the viewers who identified as LGBT experienced "the positive emotion of hope, an elevated sense of mental energy, and pathways for goals" after seeing themselves reflected in the clips they were shown, while heterosexual viewers experienced more negative responses.[33]

Finally, ABC Family/Freeform's branding and audience-building innovations have appeared in the scholarship. Louisa Ellen Stein considers the courting of fans by corporate interests in the promotion of *Kyle XY* in "'Word of Mouth on Steroids': Hailing the Millennial Media Fan." Stein identifies the fan community of *Kyle XY* as primarily interacting on the official ABC Family website rather than on fan-run sites. She concludes that fans' activities on these official sites "may not be substantively different" than activities that occur in fan-run domains.[34] Jennifer Gillan describes ABC Family's continuing efforts to build fan communities through social media in *Television Brandcasting: The Return of the Content-Promotion Hybrid*; the network mobilized *Pretty Little Liars* fans on social media by dropping clues and using social media "like a friend would," creating a loyal community of fans who boosted ratings.[35]

This volume aims to fill some of the holes in the existing scholarship surrounding ABC Family/Freeform. At this book goes to publication, ABC Family is at a critical juncture as it transforms itself into Freeform, with a new audience and new programming. This volume gives scholars the space to reflect on ABC Family's previous efforts to reach a young audience and Freeform's window of opportunity to reinvent itself as an edgier, older-skewing network that reflects the needs of today's millennial or Becomer audience.

Book Overview

Organized in four sections, this book breaks down some of the key contributions and developments that ABC Family, now Freeform, has made to the television landscape. The first section, "TV, Social Media and Fandom," examines the changing ways that people actually watch programming in the 21st century, with extended explorations of the network's innovations. By comparing two programs, *The Nine Lives of Chloe King* and *Shadowhunters*, Joe Lipsett examines how two shows, which premiered only five years apart and which garnered very similar ratings, were treated very differently by the network: cancelled and renewed for three seasons so far, respectively. Along the way, he chronicles how the definition of successful television has changed thanks to different viewing methods and modes of interactivity. In her essay, Cara Dickason continues to explore social media and its functions, though she views it through the lens of complicated teen drama *Pretty Little Liars.* Examining the way that surveillance and the gaze constantly impact the five main characters, Dickason makes connections to the way that the increased presence of social media has particularly affected young women. Through his exploration of youthful fetishism and theoretical explorations of society's capitalist desires, Stephen P. Smyth posits that *Shadowhunters* is a show focused on youth and immortality. Each of these essays provokes thoughtful consideration of how television can no longer be understood in the same way it was just ten years ago.

"Today's Feminisms" includes three essays that look at the way feminism and its different incarnations are played out across the network. Erica Lange challenges common expectations of friendships. By looking at the friend group in *Pretty Little Liars,* she argues that by having more than one best friend, women can have more fulfilling lives. By focusing on specific aspects of women's experiences, Jessica Ford dissects how labor is depicted on ABC Family. Breaking down the various types of feminism on display in *Jane by Design* (2012) and *Bunheads,* she aligns the creative labor depicted on the programs with femininity, all the while questioning if the shows take post-feminist stances. The last essay in this section, by Madeline Rislow and Anne Dotter, looks at how the rising young gymnasts in *Make It or Break It* can be seen as empowering young women in a more traditional feminist sense, breaking away from the postfeminist ideas that have gained popularity in the past thirty years. In each of these essays, ABC Family/Freeform is shown as being willing to tackle risky programming and, in particular, promote shows that encourage young women to consider their own understandings of their femininity and feminism.

In the third section, "A New Kind of Family," the authors examine the changing depictions and compositions of the family units that appear in the past ten years on ABC Family/Freeform, a topic that is especially relevant considering ABC Family's late 2000s branding efforts. First, Stephanie L. Young and Nikki Jo McCrady examine the critically acclaimed show *The Fosters,* which features lesbian parents who raise five children (a combination of a birth child, two adopted children, and two foster children). Through sensitive and thoughtful storytelling, the authors illustrate how *The Fosters* is at the forefront of progressive depictions of queer parenting. Donica O'Malley also looks at a family that incorporates all different types of people, but from an entirely different circumstance in *Switched at Birth.* In her essay, she looks at the complicated way that the characters explore race as they come to terms with the reality of growing up with the wrong families after being switched at the hospital as newborns. In the final essay of this section, Patrice A. Oppliger and Mel Medeiros look at problematic ways that foster care is often portrayed on television, yet note that in their willingness to both tackle the subject and include it so frequently, the network is actually ahead of the curve. Through diversity of their characters and inclusion of a number of themes not ordinarily addressed, ABC Family/Freeform has worked to showcase families that are not typically seen elsewhere.

Finally, "Identity Issues and 'Becomer' TV" explores the way that ABC Family/Freeform is attempting to attract audience members who are not children and teens, but rather are Becomers, in addition to the network's focus on diverse identities. Sharon L. Pajka looks at the way that the creators of *Switched at Birth* have explored Deaf culture. While they have made much progress by including Deaf actors and themes, Pajka makes clear there is still more work to be done in the area of equitable representation. Similarly, Malynnda A. Johnson and Kathleen M. Turner argue that television programs, and the network itself, need to address sexual health topics more, particularly when it comes to sexually transmitted infections. Yet, at the same time, they draw attention to the idea that more than ever before, ABC Family/Freeform has worked to create better conversations and establish more realistic portrayals of sexual relationships between young people in their programming.

Pursuing similar interests, the last two essays in this section are focused on how identity is constructed and defined in ABC Family/Freeform's programming. Andi McClanahan does an in-depth reading of the show *Huge* (2010), which never found its audience and was cancelled after one season. By continuing to problematize and shame fat bodies, a show that marketed itself as body-positive failed to achieve those goals. Lastly, Anelise Farris shows that not only did *Switched at Birth* include positive views of disabled characters, the show's creators worked to create complicated, nuanced plotlines. In that sense, the show has the potential to educate its viewers, helping them to see and understand the lives of the disabled characters on the screen and hopefully lead to a more accepting society.

Freeform carries on the legacy that ABC Family began. Considering topics that other networks steer away from, Freeform not only tackles them but considers their impact on young adults. A 2017 promo video for *The Fosters*, "Family Stories," features the show's actors reading letters and feedback from viewers of the program. Many of the actors tear up, as the fans have shared how much this show matters to them. Sherri Saum, the actress who plays Lena Adams Foster, pointedly observes why people are responding so powerfully to the show, saying that she is proud to be on the program that "gives representation to people that don't have that representation."[36] While the actors may be emotional, it is clear that viewing the show has strongly affected viewers, as they share how grateful they are to see stories about homosexuality, transgender issues, foster care, and adoption. Powerfully, one audience member has written, "After viewing the show, I learned to accept those who are different."

This book considers ABC Family/Freeform's groundbreaking attempts to consider an important television-watching audience. From teenagers to young adults to adults in their early thirties, the network not only takes this demographic seriously as an audience and as consumers, but they also incorporate programming and themes that have not often been addressed by others. Freeform makes the case that we should not dismiss young people and their concerns too lightly; rather, the network emphasizes that the Becomers are some of the most progressive, challenging people consuming media today. Smartly, Freeform gives Becomers a voice, thereby recognizing that they are not just the leaders of tomorrow, but they are the ones who change society, and as many of these essays demonstrate, Becomers are demanding that the world become a more diverse and accepting environment.

NOTES

1. Lesley Goldberg, "ABC Family Leaves the Nest and Changes Its Name," *Hollywood Reporter*, October 16, 2015, 10.
2. *The 700 Club* has been broadcast continuously from 1966 to present. It began on a Virginia affiliate station before being broadcast on the new Christian Broadcasting Network (CBN) in 1977. It has remained on the network that is now Freeform through all of its name and ownership changes: CBN Cable Network (1977–1988), The Family Channel (1988–1998), Fox Family (1998–2001), ABC Family (2001–2016), and Freeform (2016–present).
3. Caryn Murphy, "Secrets and Lies: Gender and Generation in the ABC Family Brand," in *The Millenials on Film and Television: Essays on the Politics of Popular Culture*, ed. Betty Kaklamanidou and Margaret Tally (Jefferson, NC: McFarland, 2014), 18.
4. Louisa Ellen Stein, *Millennial Fandom: Television Audiences in the Transmedia Age* (Iowa City: University of Iowa Press, 2015), 16.
5. The history of this channel, from its CBN beginnings to its connection to ABC, is extrapolated from the detailed essay by Caryn Murphy, "Secrets and Lies."
6. Julie Liesse, "Getting to Know the Millennials," *Advertising Age*, July 9, 2007, A1–A6.
7. Stein, *Millennial Fandom*, 16.
8. Brooks Barnes, "ABC Family Spins Gold in Dramas," *The New York Times*, March 6, 2011.
9. Goldberg, "ABC Family Leaves the Nest."
10. ABC Family's transformation into Freeform provided an opportunity for other, smaller networks to step in and provide the family-oriented programming that ABC Family had focused on in its early years. UP Network, for example, has found its audience by airing *Gilmore Girls*, *7th Heaven*, and other WB network shows that had previously been seen on ABC Family. See Mike Farrell, "The Struggle for Independents," *Broadcasting & Cable*, March 6–13, 2007.
11. For more about its dominance, see Josef Adalian, "Why *Pretty Little Liars* Was Such a Crucial Hit for Freeform," *Vulture*, June 29, 2017, http://www.vulture.com/2017/06/pretty-little-liars-ratings-finale-freeform.html.
12. Neal Justin, "Social-Media Hit *Pretty Little Liars* Rode Influence to Ratings Success,"

Minneapolis Star Tribune, June 26, 2017, https://www.thestar.com/entertainment/television/2017/06/26/social-media-hit-pretty-little-liars-rode-influence-to-ratings-success.html.

13. Qtd. in Lesley Goldberg, "Marvel's 'New Warriors,' with Squirrel Girl, Ordered Straight-to-Series at Freeform," *Hollywood Reporter*, April 15, 2017, http://www.hollywoodreporter.com/live-feed/squirrel-girl-tv-series-freeform-orders-marvels-new-warriors-991277.

14. Qtd. in Scott Porch, "'The Bold Type,' a Modern Gloss on Sex and the Single Girl," *The New York Times*, June 30, 2017, https://www.nytimes.com/2017/06/30/arts/television/the-bold-type-a-modern-gloss-on-sex-and-the-single-girl.html?_r=0.

15. Erik Pederson, "Freeform Trailers: 'Marvel's Cloak & Dagger,' Lonely Island's 'Alone Together' & Drama 'The Bold Type'—Upfront," *Deadline*, April 19, 2017, http://deadline.com/2017/04/cloak-and-dagger-trailer-alone-together-the-bold-type-freeform-upfront-1202072282/.

16. Leslie Goldberg, "Freeform's Fraught Year: When a Channel Chases Millennials," *The Hollywood Reporter* 35, November 18, 2016.

17. Stein, *Millennial Fandom*, 17.

18. *Ibid.*, 18.

19. Murphy, "Secrets and Lies," 26.

20. *Ibid.*, 24.

21. Roland T. Rust and Naveen Donthu, "A Programming and Positioning Strategy for Cable Television Networks," *Journal of Advertising* 17, no. 4 (1988): 6.

22. Heather Hundley, "The Evolution of Gendercasting: The Lifetime Television Network—'Television for Women,'" *Journal of Popular Film and Television* 29, no. 4 (2002): 180.

23. Amanda Lotz, *Redesigning Women: Television After the Network Era* (Urbana: University of Illinois Press, 2006).

24. *Ibid.*, 6.

25. *Ibid.*, 176.

26. Amanda Lotz, *The Television Will Be Revolutionized*, 2d ed. (New York: New York University Press, 2014), 8.

27. *Ibid.*, 267–268.

28. Melanie E.S. Kohnen, "Cultural Diversity as Brand Management in Cable Television," *Media Industries Journal* 2, no. 2 (2015): 88–103.

29. Jennifer Mitchell, "The Best Lesbian Show Ever! The Contemporary Evolution of Teen Coming-Out Narratives," *Journal of Lesbian Studies* 19, no. 4 (2015): 454–469.

30. Wendy Peters, "Bullies and Blackmail: Finding Homophobia in the Closet on Teen TV," *Sexuality and Culture* 20, no. 3 (2016): 486–503.

31. Shane Brown, "Pushing Boundaries? Challenging Traditional Gender Roles in *Kyle XY*," *Science Fiction Film and Television* 8, no. 1 (2015): 91–100.

32. Seon-Kyoung An, Llewyn Elise Paine, Jamie Nichole McNiel, Amy Rask, Jourdan Taylor Holder, and Duane Varan, "Prominent Messages in Television Drama *Switched at Birth* Promote Attitude Change Toward Deafness," *Mass Communication and Society* 17, no. 2 (2014): 195–216.

33. Traci K. Gillig and Sheila T. Murphy, "Fostering Support for LGBTQ Youth? The Effects of a Gay Adolescent Media Portrayal on Young Viewers," *International Journal of Communication* 10 (2016): 3828–3850.

34. Louisa Ellen Stein, "'Word of Mouth on Steroids': Hailing the Millennial Media Fan," in *Flow TV: Television in the Age of Media Convergence*, ed. Michael Kackman, Marnie Binfield, Matthew Thomas Payne, Allison Perlman, and Bryan Sebok (New York: Routledge, 2010), 128–143.

35. Jennifer Gillan, *Television Brandcasting: The Return of the Content-Promotion Hybrid* (New York: Routledge, 2015), 255–259.

36. Freeform, "The Fosters: Family Stories," *YouTube*, July 6, 2017, https://www.youtube.com/watch?v=9z4vCJyZw-M&index=30&list=PLt1alZH4ufeoqIisSgpC5M8SLSri91DpZ.

Defining Success in the Era of Peak TV

A Case Study of The Nine Lives of Chloe King *on ABC Family and* Shadowhunters *on Freeform*

JOE LIPSETT

Shadowhunters: The Mortal Instruments (2016–present, hereafter known as *Shadowhunters*) was the first television series to debut on Freeform after the network changed its name from ABC Family in January 2016. The timing is significant: *Shadowhunters*, an adaptation of the successful young adult (YA) book series by Cassandra Clare, is Freeform's first supernatural fantasy series. ABC Family previously dabbled in the genre with *Kyle XY* (2006–2009), but *Shadowhunters'* closest comparison series is *The Nine Lives of Chloe King* (hereafter known as *Chloe King*), which debuted in 2011. Both *Shadowhunters* and *Chloe King* are adaptions of YA book series and feature teen girl protagonists who discover that they are members of a super-powered species caught up in a centuries-long battle against a warring enemy. Despite similar ratings and generally complimentary to negative reviews, ABC Family and Freeform treated the two series in markedly different ways: *Chloe King* was cancelled following its first season, while *Shadowhunters* was renewed through season three.

This essay explores how the definition of a successful television series evolved in the five years separating *Chloe King*'s cancellation on ABC Family and *Shadowhunters'* premiere on Freeform. The changing landscape of contemporary television, marked by media and audience fragmentation, has resulted in a rise in the number of television networks and scripted television series in the last decade. This phenomenon—coined Peak TV by FX Networks President John Landgraf at a Television Critics Association (TCA) presentation in 2015[1]—prompted ABC Family not only to rebrand itself as Freeform, but to shift its target audience and, most significantly, adopt the social media strategies perfected by flagship series *Pretty Little Liars* (2010–2017). In the time that separates the debut of *Chloe King* and *Shadowhunters*, the determinants of success on ABC Family/Freeform (and by extension most television networks) have shifted from a sole focus on live ratings to including data on streaming viewers on digital platforms, as well as audience reach and engagement on social media platforms such as Facebook and Twitter before, during, and after new episodes.

Peak TV: Scripted Television in the 2010s

The television landscape has changed dramatically in the last decade. There is increased competition among networks for viewers, which has seen traditional broadcast networks such as NBC, ABC, CBS, FOX, and The CW compete with premium cable channels such as HBO, FX, AMC, Showtime, and STARZ, as well as online streaming giants like Netflix, Hulu, and Amazon. The rapid proliferation of TV series in the era of Peak TV—what television critics[2] consider both a blessing and a curse—is the new reality of the industry. In 2015, 409 scripted series were produced—nearly doubling the number from 2009.[3] In 2016 that number rose to 455, with speculation that the number could potentially hit 500 series in 2017.[4]

As part of his upfront presentation at the 2015 TCAs, Landgraf shared with television critics a bar graph from FX Networks Research visualizing this dramatic spike in scripted television from 2002 to the present. Of note for this discussion is the increase in the number of original scripted series between 2011 (the year that *Chloe King* was released on ABC Family) and 2016 (the year that *Shadowhunters* debuted on Freeform): 266 to 455. This marks a 71 percent increase in the number of scripted television series vying for audience attention in only five years. During this time ABC Family underwent a fairly significant change of its own: it became Freeform.

ABC Family's Brand Identity

ABC Family has undergone two significant rebrands in the twenty-first century. The first occurred in July 2001 when Fox Family was sold to The Walt Disney Company and renamed ABC Family.[5] For its first few years, the newly renamed network struggled to establish its identity. Issues with the ownership of syndication rights prevented the channel from airing anything other than series produced by corporate parent studios Walt Disney Television and Touchstone Television[6] or second-run series purchased from the WB network, whose programming explicitly targeted teen female audiences.[7] This situation did not change until 2006 when the network adopted a new tagline and began programming its own drama and comedy series. Caryn Murphy dedicates a substantial portion of her essay "Secret and Lies: Gender and Generation in the ABC Family Brand" to ABC Family's WB–like programming and branding practices, which included hiring WB talent (Senior VP Development Kate Juergens), series creators (Amy Sherman-Palladino, *Gilmore Girls* [2000–2007]; Brenda Hampton, executive producer of *7th Heaven* [1996–2007]) and production partners (Alloy Entertainment).[8] ABC Family branded their original programming to a specific target audience: millennial viewers, the "83 million Americans born between 1977 and 1996" who "spend almost 15 hours a day interacting with various media and communications technologies."[9] This audience was explicitly named in a 2007 ABC Family advertisement published in *Advertising Age* entitled "Getting to Know the Millennials"; it serves as an example of the network's attempt to target a specific, niche audience, which is one of the strategies that conglomerates have increasingly adopted in the "neo network," digital era.[10]

ABC Family was programming for a specific audience: 12- to 24-year-old "tech-savvy, second-screen focused [females] … longing for connectivity,"[11] and in this regard they were successful. In 2015, the final year before the network changed its name, ABC Family not only had its most competitive year in the ratings among women 18–49, it was the second most popular cable network for females 12–34.[12]

The perception of the network by non-viewers was a different story. ABC Family network president Tom Ascheim told reporters that "recent research conducted by the network [revealed] that non-viewers identified the network as 'family-friendly' and 'wholesome' and gave 'pretty lousy marks on everything else.'"[13] The network's name proved to be a limiting factor; non-viewers associated it too much with family—a "perception gap" and a "hindrance" to growth.[14] Ascheim and the other network executives deemed the perception problems severe enough to justify a second major rebrand, and in October 2015 Ascheim announced that ABC Family would be known as Freeform starting January 2016. He explained that the new name reflected how "media is oozing onto all sorts of different screens and platforms and it feels kind of free form in the way that people find it."[15]

The name change redefined the network's target audience: millennials became "Becomers"—a term created by Freeform to describe the new generation of youths and young adults aged 12–34, or the time span between "first kiss and first kid."[16] Ascheim explained: "We think that they are indeed in formation so on some level, so it really feels like it invokes the quality of being a Becomer."[17] While the term Becomer was initially introduced in an upfront presentation in April 2015, by the time the network name change was officially announced in October, the network's priority had shifted to focus on audience engagement online, specifically the opportunity to capitalize on "users" as participatory, "active contributors."[18] ABC Family's Senior Vice President of Marketing, Creative and Branding, Nigel Cox-Hagan, made this sentiment explicit: "As the most social cable network, the channel already has a captive audience readily awaiting information to be tweeted at them, not to mention highly socially engaged talent."[19]

The shift in priorities is clear. When ABC Family presented the results of their research in the *Advertising Age* article on millennials in 2007, the network was interested in attracting tech-savvy audiences. Fast-forward eight years and Freeform is no longer simply content with viewers who watch content on second-screens; the rebranded network wants audiences who see themselves as content collaborators and active participants. This is the new normal in the age of Peak TV: as audience attention is fragmented across hundreds of series, live ratings (captured the same day an episode airs and represented in audience by millions that are broken down into key demographic metrics) no longer accurately capture how or when audiences engage with their favorite series. Tools and platforms that were in their infancy in 2011 evolved or paved the way for broader, more accessible options for audiences to get involved. Viewers, as a result, no longer simply watch programs; they actively consume, engage and generate content about their favorite series—activities that networks have used to supplement or complement traditional live ratings to demonstrate viewer interest in their programming (effectively monetizing social media engagement to advertisers). The evolution of audience measurement from live ratings to engaged, participatory, collaborative audiences is not unique to ABC Family/Freeform—it applies to an entire television industry struggling to transition between traditional and new methods for measuring television audiences in the era of Peak TV when live ratings have fallen across the board. This shift is key to understanding how the definition of success for original scripted television series has changed in the intervening years between the 2011 premiere of *Chloe King* on ABC Family and arrival of *Shadowhunters* in 2016 at Freeform.

Case Study: The Nine Lives of Chloe King *vs* Shadowhunters: The Mortal Instruments

The striking similarities between *Chloe King* and *Shadowhunters* make them an exemplary pair for highlighting how the industry has changed in five years. Both series are adaptations of YA book series. The decision to greenlight films and television based on pre-existing properties is a well-established economic model for Hollywood and has become increasingly popular as a result of escalating production costs and the desire to reduce financial risk by capitalizing on built-in fanbases.[20] This practice extends to teen-oriented television series made in the 2010s, many of which began as female-fronted YA book series. While there is a tradition of lumping all teen programming into a single overarching genre dubbed Teen Television,[21] there remain unique genre conventions and iconographies that appeal (and can therefore be marketed) to different audiences.[22] For the most part, YA book series adapted into teen television series can be separated into the broad genres of Melodrama/Romance, Mystery/Thriller and Science Fiction/Supernatural.

- *Melodrama/Romance* adaptations are typically set in the "real world," and conflict is predominantly the result of romance and relationship drama. These series are frequently set in large metropolitan cities and feature middle class to wealthy characters. A representative example is *Gossip Girl* (The CW, 2007–2012).
- *Mystery/Thriller* series are more likely to be set in small towns or the suburbs. While there is still conflict arising from romance and relationship drama, it is juxtaposed or complicated by crime, lies, and secrets. There is often a mystery or a threat that puts characters in jeopardy and must be solved. Representative examples include *Pretty Little Liars* or *13 Reasons Why* (Netflix, 2017–present).
- *Science Fiction/Supernatural* series are distinct from other teen genres in that they are more likely to be set in the future, alternate worlds or feature supernatural elements. Although conflict still arises as a result of romance and relationship drama, these series create drama from the intersection of the real world and science fiction or the supernatural. A representative example is *The Vampire Diaries* (The CW, 2009–2017).

The following table breaks down recent teen television adaptations of YA books by genre. Each column includes the details of the TV adaptation, including the number of years the television series ran and on which network, as well as the name of the novel(s), the name of the author, the dates when the novel or series were published and, where applicable, the number of books in the series.

The table illustrates that teen television series adapted from YA series not only fall into similar genres, but that they are predominantly spread out across three networks: The CW (formerly The WB and UPN), MTV, and ABC Family/Freeform. Considering The WB's history of developing series for teen girl audiences, it is unsurprising that the network has a large number of YA series adaptations. Likewise, MTV has always catered to teen audiences,

Genre	Television Series	Network	YA Book Series	Number of Books
Melodrama/ Romance	*Gossip Girl* (2007–2012)	The CW	*Gossip Girl* (Cecily von Ziegesar, 2002–2013)	13
	The Carrie Diaries (2013–2014)	The CW	*The Carrie Diaries* (Candace Bushnell, 2010–2011)	2
Mystery/ Thriller	*13 Reasons Why* (2017–present)	Netflix	*Th1rteen R3asons Why* (Jay Asher, 2007)	1
	Pretty Little Liars (2010–2017)	ABC Family/ Freeform	*Pretty Little Liars* (Sara Shepard, 2006–2013)	14
	Eye Candy (2015)	MTV	*Eye Candy* (R.L. Stine, 2004)	1
	The Lying Game (2011–2013)	ABC Family	*The Lying Game* (Sara Shepard, 2010–2013)	6
Fantasy/ Supernatural	*The Shannara Chronicles* (2016–present)	MTV	*The Shannara Chronicles* (Terry Brooks, 1977–1985)	3
	The 100 (2014–present)	The CW	*The 100* (Kass Morgan, 2013–2016)	4
	The Vampire Diaries (2009–2017)	The CW	*The Vampire Diaries* (L.J. Smith, 1991);	4 (original)
			The Return (2009–2011);	3
			Stefan's Diaries (2010–2012);	6
			The Hunters (2011–2012);	3
			The Salvation (2013–2014)	3
	The Secret Circle (2011–2012)	The CW	*The Secret Circle* (L.J. Smith, 1992)	3

first with music videos and then increasingly with scripted fare including, most recently, an adaptation of Terry Brooks' fantasy series *The Shannara Chronicles* (2016; moving to the newly renamed Paramount Network in 2017 for season 2). The final network is ABC Family/Freeform, which has adapted several series by successful author Sara Shepard, the creative mind behind both *Pretty Little Liars* and *The Lying Game*.

Considering the move towards edgier fare that began with *Pretty Little Liars*, it is unsurprising that ABC Family choose to adapt *Chloe King* and *Shadowhunters*. Both series are strategically on brand for the network: they feature female protagonists within the desirable age demographic and story lines that provide plenty of opportunities for audience-relatable romance and relationship drama. The content of both series also reflects ABC Family's appeal to greater diversity in race, ethnicity, and sexual orientation,[23] and the conventions of the supernatural genre distinguish *Chloe King* and *Shadowhunters* from the network's other series while still providing a sexy, mysterious hook to entice viewers and draw in fans of their respective source material.

The Nine Lives of Chloe King

The Nine Lives of Chloe King is based on a trilogy of books written by British author Liz Braswell, under the pseudonym Celia Thomson. The first two volumes, *The Fallen* and *The Stolen*, were both published in 2004, with the third volume, *The Chosen*, arriving the following year. Both the books and the television series follow the titular character, Chloe King, who develops unusual cat-like powers on her sixteenth birthday. Chloe soon learns that she is a member of an ancient race called the Mai, which has been hunted by human assassins throughout history—and that she may be the long-prophesized savior who can bring an end to the war.[24]

The ten episode first season was announced January 31, 2011, as part of then–ABC Family President Michael Riley's first pick-ups.[25] The series debuted Tuesday, June 14, following the premiere of *Pretty Little Liars'* second season. Critics wrote generally favorable reviews (the series scored 68/100 from 9 critics and a 7.8/10 user score by 30 site users on review website Metacritic), though the pilot was watched by 2.2 million viewers, a figure well below lead-in *Pretty Little Liars* (3.7 million) and fellow new series *Switched at Birth* (3.3 million), which premiered earlier that month.[26] Over the remainder of the season, *Chloe King* lost half of its audience, with the finale delivering a 0.3 18–49 rating and 1.078 million viewers.[27] News of the cancellation was reported September 15, 2011, with the suggestion that "soft ratings"[28] had doomed the series.

Shadowhunters: The Mortal Instruments

Shadowhunters is based on a series of six books entitled *The Mortal Instruments* written by Cassandra Clare from 2007 to 2014. The official website for the *Shadowhunters* book series lists the sales for Clare's novels at 36 million worldwide,[29] confirming that it is extremely popular among readers. *Shadowhunters* follows 18-year-old Clary Fray, who finds out on her birthday that she is a Shadowhunter, one of a group of human-angel hybrids who hunt demons. Not unlike *Chloe King*, the revelation about her real identity introduces Clary to a new world filled with supernatural creatures while simultaneously positioning her as the missing ingredient that will tip the balance of power in the conflict.[30]

Prior to its television incarnation, the first novel was adapted as a feature film, *The Mortal Instruments: City of Bones*, in 2013. The film was produced by Constantin Film and Unique Features and starred a British cast of two relatively unknown young actors, Lily Collins and Jamie Campbell Bower, surrounded by more-established character actors Lena Headey, Aidan Turner, Jared Harris, and Jonathan Rhys Meyer. While the film was critically derided (12 percent fresh on *RottenTomatoes*, B+ Cinemascore from audiences), it did connect with its primary audience: BoxOfficeMojo.com reported that "the movie's audience was 68 percent female and 46 percent under the age of 21."[31] Despite a $59 million gross outside of North America, the $60 million film only grossed $31 million in the United States. While the poor reviews likely didn't help, audience fatigue may have also affected the box office success of the film. *The Mortal Instruments* was the third YA adaptation to hit theaters in 2013, following *Beautiful Creatures* and *The Host*, both of which "bombed" with sub-$30 million grosses.[32] Plans for a sequel to *The Mortal Instruments* were subsequently scrapped.

Despite the failure of the film, tapping into a property with a preexisting fan base of 36 million was too tempting to resist, particularly in a television climate that has seen a substantial spike in the number of adaptations of well-established properties in the last few years.[33] In October 2014, Constantin Films announced Clare's property was being developed as a television series under showrunner Ed Decter.[34] It was picked up to series by ABC Family in late March 2015,[35] at which time network president Tom Ascheim directly referenced the books' popularity alongside other established YA franchises: "*Shadowhunters* is a big epic saga that will resonate with viewers who come to ABC Family for the *Harry Potter, Hunger Games*, and *Twilight* franchises. A *New York Times* bestseller for 122 consecutive weeks, with over 35 million copies in print worldwide, *Shadowhunters* is the perfect story to share with our audience."[36] The reference to other successful adaptations implies that *Shadowhunters* was greenlit thanks to its preexisting, built-in audience, and because it was perceived as a good fit for the network's other programming. Four days before the show's January debut, then-showrunner Decter attended a TCA panel with members of the cast, indirectly describing how the adaptation fit into Freeform's rebranding strategy. Decter explained that "the coolest thing" was how the show matched the rebrand of the network. He continued, "I said the show is gonna be darker, sexier, it's got some racy content, and they said bring it on."[37] The series premiered on Tuesday, January 13, 2016, the night of Freeform's official launch, in the 9 p.m. timeslot following the season six winter debut of *Pretty Little Liars*. Significantly, *Shadowhunters* premiered in the same timeslot as *Chloe King*. In its debut, *Shadowhunters* attracted 1.82 million viewers and a 0.8 rating of the coveted adults 18–49 demographic[38]—a lower figure than *Chloe King*.

The darker, sexier, racier Freeform series didn't fare much better with critics than the film adaptation did, scoring a mixed to average score of 45/100 from 9 critics and a generally favorable 6.3/10 from Metacritic users. Much like *Chloe King*, the supernatural series lost a substantial portion of its audience over its first season, ending its 13 episode run on April 5, 2016, with 0.8 million viewers, nearly 300,000 fewer viewers than *Chloe King* did five years earlier. So why did *Shadowhunters* survive when *Chloe King* was cancelled? Simply put: social media.

The Rise of Social TV

The manner in which television is consumed over and across multiple platforms continues to dominate the scholarly discussion about media. In the era of Peak TV, one of the greatest challenges for networks is audience fragmentation, wherein potential viewers are dispersed across an unprecedented array of networks and content options—in this case, hundreds of networks airing nearly 500 television series.[39] It is increasingly common for television series to seek out ways to engage audiences in order to retain their attention, generate buzz and—now that live ratings are no longer the sole method of documenting the size of the audience to advertisers—create a social TV narrative that can be monetized.

This drive for a strong social media presence is important in two regards: first, social media generates viewership, loyalty, and engagement in a series before, during, and following episodes courtesy of fan-created "organic content," and second, social media generates revenue by driving traffic to official ad-supported websites and outlets using "owned content"—social content that is pushed out by networks in the form of clips, extras, and games.[40] As audience fragmentation shrinks the size of live television viewing audiences, social media engagement (comments, likes, retweets) and reach (who sees the content) of both owned and organic content, broken down by demographic, is increasingly important for networks seeking to partner with marketers looking for access to their desired consumer base.[41]

One method is to adopt what Henry Jenkins calls a "transmedia" storytelling approach, which is defined as "a process where integral elements of a fiction get dispersed systematically across multiple delivery channels for the purpose of creating a unified and coordinated entertainment experience."[42] There is a history of teen television series adopting transmedia strategies, including the use of media created specifically for online platforms to complement episodes, harness fan activity, and sustain interest. Examples include an interactive website called "Dawson's Desk" for *Dawson's Creek* (The WB, 1998–2003) and "The O.C. Insider," an exclusive fan site created for *The O.C.* (FOX, 2003–2007).[43] These approaches are hardly novel: Usenet and discussion forums were popular platforms for fans to connect, speculate, and digest the latest episodes of several television series in the 1990s, particularly cult series such as *Twin Peaks* (ABC, 1990–1991), *The X-Files* (FOX, 1993–2002) and *Buffy the Vampire Slayer* (The WB, 1997–2001; UPN, 2001–2003).[44]

Ryan Cassella argues that following the arrival of cable and satellite content providers, television has entered a new critical "wave": "a new media evolution" of connectivity that has given the user unparalleled choice in their entertainment options," the span of which has never been "more diverse or engaging."[45] These new methods of accessing and consuming television are reflective of the media landscape that evolved and solidified following the premiere of *Chloe King*. Although second-screen viewing existed in 2011, it was less pervasive, and live viewing was still considered the dominant mode of evaluating success. Even in 2011, media and audience fragmentation was creating reliability issues with audience measurement data, which struggled to accurately capture viewing figures as new alternatives began to gain traction.[46] In 2017, it remains desirable to have strong live ratings, but there is now a begrudging acknowledgment that live ratings have slipped uniformly across the board as more and more audiences seek out flexible viewing alternatives, such as mobile and streaming options.

When *Chloe King* debuted, these flexible viewing options were only just beginning to appear, and there was significant difficulty in accurately measuring them.[47] Today, there is greater consistency in the measurement of delayed viewing (such as DVR) ratings, as well as streaming video on demand (VOD) via iTunes, Netflix, and network websites and apps. It is common for series to be available on digital platforms the same day they air on television. *Shadowhunters*, for example, was available on the Freeform website, Hulu, and Netflix Canada from its very first episode. And while the live ratings of *Shadowhunters* and the other Freeform series fell below the live ratings of other ad-supported networks that subsist on advertising revenue in 2016, if non-linear viewership on platforms such as the Freeform app, Hulu, and VOD are factored in over a 35-day window, Freeform series add an average of 33 percent to their adult 18–34 ratings demographics. Additionally, total consumption across all digital platforms for 2016 was up 47 percent, from

4.9 billion minutes to 7.2 billion.[48] The viewing landscape for Freeform's audience has clearly shifted towards flexible viewing alternatives.

Statistics from digital platforms are not the sole metric that networks take into consideration when evaluating the success of their programs, however, particularly one as young-skewing as Freeform. The measurement of Social TV, social media engagement, has become increasingly important to network executives and advertisers in the last five years.[49] A significant component of the "new media evolution" is the role social media plays as the "middle-man" between creator and consumer[50] in helping to create a "comprehensive media experience" wherein the television series is but one component.[51] Market researchers believe that television viewing and social media are increasingly being consumed simultaneously, particularly when there are explicit on-screen prompts such as hashtags.[52] Mike Proulx and Stacey Shepatin, authors of *Social TV*, argue that "social media creates the feeling of a more direct and intimate interaction between celebrities and their fans" leading to a "more engaged audience."[53] Currently, it is rare to find a television series that does not produce regular content across a variety of social media platforms, particularly on young-skewing networks such as The CW, MTV, and Freeform. As media fragmentation increases, especially over the last five years with the rise of streaming players such as Amazon, Hulu, and Netflix, the need for networks to communicate and engage with audiences in this fashion has become crucial to their success.[54]

Social media leaders in television identify Facebook and Twitter as the two dominant social media platforms that television networks use to engage fans.[55] Facebook is best used for "long-term campaigns with multimedia content and engaging fans between program airings," while Twitter is more of a "TV-friendly interface," noted for its immediacy and responsiveness for "capturing audience feedback and engagement."[56] As Murphy, Ariane Galope, and Melanie Brozek all argue, ABC Family/Freeform has been incredibly successful at using social media platforms to engage with their target millennial/Becomer audience.

In this regard, *Pretty Little Liars* remains the gold standard. As ABC Family/Freeform's most successful series, *Pretty Little Liars* commands a massive social media footprint. Although live ratings have declined in recent years, in 2016 the series was still the dominant television series on social media with a social engagement score of more than 256 million interactions.[57] In April 2017, *Pretty Little Liars* maintained nearly 4 million Twitter followers and 14.7 million Facebook fans.[58] This translates into a captive audience of millions who are regularly exposed to product-placement, ad-supported videos, and who are redirected to sponsor and network websites.

Television social media leaders espouse the merits of using social networks in three different timeframes: before, during, and after a program.[59]

One strategy for using social media platforms *before* a new episode is the creation of "event television," putting social channels and audiences to work generating free advertising before the program airs.[60] Brozek dedicates part of her analysis of *Pretty Little Liars* to this phenomenon, exploring how contests and prompts are used by the show's official Twitter handle to give fans a "cause" to engage with the series and get it trending.[61] There is evidence that the official *Shadowhunters* Facebook (ShadowhuntersSeries) and Twitter handle (@Shadowhunters) also use these techniques, prompting fans to respond to questions that are specific to individual episodes (a 30-second promo posted on Facebook eight hours before "Dust and Shadows" [season 2, episode 5, aired on January 30, 2017] included the tagline "Can Clary bring her mom back from the dead?") and posing broader, general questions about fans' interests and engagement with the series (one January 28, 2017, midweek tweet simply asked "What's YOUR favorite rune?," referencing the characters' tattoos).

This differs from the use of social media *during* a program, which is more focused on creating a community via a shared real-time experience, thereby making the audience part of the show, rather than maintaining their engagement between episodes.[62] Networks have the capacity to embed and encourage audience interaction directly into the program with onscreen hashtags during key, marketable moments of a telecast. This can result in a dramatic increase in the number of backchannel tweets, or tweets that occur simultaneously during a live event, allowing passive viewers to become active participants by sharing comments, questions, and criticisms.[63] This strategy occurred during *Shadowhunters'* second season premiere on January 2, 2017, when an in-episode advertisement encouraged audiences to "Live Tweet Now! Chat Now with the Stars from Shadowhunters @ShadowhuntersTV #ShadhowhuntersChat"—a unique opportunity for fans to engage not only with the show, but with the cast members themselves, in real time.

Live-tweeting by cast members is another dedicated way to engage fans online. Research conducted by Twitter in 2014 reinforces the value of having talent involved on social media platforms and developing relationships with fans, particularly while episodes air. Anjali Midha, the former Global Media & Agency Research Director for Twitter, confirms that cast members who live-tweet during a show's premiere generate 64 percent more Tweets that day than shows that do not. Midha positively correlates live-tweeting with audience building on the social network, stating that with live-tweeting, an official show handle can expect a "lift" (or increase) in followers of seven and a half percent, while individual cast members can see an astonishing 228 percent increase.[64]

This latter argument is especially important for science fiction series. Included in Midha's report is a graph that confirms that science fiction series with casts that live-tweet see an increase in social media presence of 218 percent over series with casts that do not live-tweet. The cast of *Shadowhunters* have clearly taken this approach to heart by adopting the social media strategies of their *Pretty Little Liars* brethren: all nine of *Shadowhunters'* second season series regulars (Katherine McNamara, Dominic Sherwood, Matthew Daddario, Alberto Rosende, Isaiah Mustafa, Harry Shum, Jr., Emeraude Toubia, Alan Van Sprang, and Maxim Roy) are active on Twitter in support of the series. As of the end of April 2017, they have a combined 4.39 million followers (Harry Shum, Jr., alone accounts for 1.21 million). Nearly all of the cast members participate each week in #ShadowhuntersChat, live-tweeting as new episodes air, replying to fans' messages, and retweeting each other and content from the show's official handle. Again, this is common practice for current television series, but these practices were uncommon back in 2011 when *Chloe King* was airing original episodes. While several of *Chloe King*'s stars were active on Twitter during the show's run, they did not live-tweet during any of the episodes.

The final component is *post-show* social media activity, which helps to bridge the gap between new episodes and keep the conversation going. The official Twitter handle for *Shadowhunters* routinely posts owned content such as promotional stills for upcoming episodes, recap videos, GIFs, behind-the-scenes footage from recent episodes, and occasionally organic content produced by fans, such as @ShumDarioNews' weekly fan reviews. There is also a frequent emphasis on engaging fans by asking for their unique, individual responses (reinforced via capitalization), such as a January 27, 2017, tweet—halfway between new episodes—that read "#FlashbackFriday to #ShadowhuntersSeason2 episode 2. What was YOUR favorite scene?"

The use of social media strategies before, during, and after episodes have proven remarkably successful for *Pretty Little Liars*, *Shadowhunters*, and other ABC Family/Freeform series. In 2015, the network ranked first for the second-straight year among women 18–34 with a combined 62 million fans and 10.6 million average engagements per show.[65] *Shadowhunters* is a significant contributor to these figures. While Goldberg's observation about falling live ratings are applicable to *Shadowhunters* (*Variety* described its season one ratings as "modest,"[66] and according to TVSeriesFinale.com,[67] the series' live viewership is down 11.8 percent in total viewers and 24 percent in the 18–49 demo year to year—second lowest among all Freeform series), it remains a dominant social media presence for the network.

For the 2017 television season, *Shadowhunters* is the fourth most engaging scripted series overall across all TV with over 17.7 million engagements,"[68] which is a consolidated measure of public engagement that includes fan growth, content responses, and conversation volume across Facebook, Instagram, and Tumblr. ListenFirstMedia, a research firm that analyzes engagement across major social media platforms, crowned *Shadowhunters* the #1 most social new series of 2016 with "an engagement score just north of 60 million."[69] When the series was renewed for season two on March 14, 2016, Freeform even used the announcement to herald the series' social media prowess, confirming that the series is performing well on a metric that the network factors into their renewal decisions. The third season renewal on Friday, April 21, 2017, made this connection even more explicit: rather than have a Freeform executive make the announcement at the TCA or via press release, the pickup was announced by the cast themselves during a live Facebook chat with fans.[70]

Conclusion

If the timeline for *Chloe King* and *Shadowhunters* were reversed, it is entirely possible that *Shadowhunters* would have been cancelled and *Chloe King* renewed. Back in 2011 when the television landscape had roughly 50 percent fewer scripted series, live ratings were the dominant form of measuring the success of television series. This is not to say that *Chloe King* did not have a social media presence: in 2017, the show's official Twitter handle still boasts 15.6 thousand followers, despite not tweeting since September 13, 2013. More impressive is the series' Facebook group, which has grown exponentially in the last few years. Using the Wayback Machine Internet Archive,[71] it is possible to determine that *Chloe King*'s Facebook account had 19,112 fans in May 2011 (before the show's official debut), a figure which has grown to 326,566 in April 2017, nearly five and a half years after the series' demise. As recently as 2016, when Freeform used the Facebook sites of cancelled series to advertise its new app, there remained a rapt, attentive (and occasionally angry) group of fans clamoring for new episodes of *Chloe King*. Additionally, both the number of likes and comments on the *Chloe King* Facebook page increased each month throughout 2016.

This essay has demonstrated that the methods for defining success for series has changed dramatically over the five-year period between *The Nine Lives of Chloe King* and *Shadowhunters: The Mortal Instruments*. The two adaptations share a genre, premise, target audience, timeslot, and lead-in, but there is a fundamental difference in how success was measured on ABC Family in 2011 compared to Freeform in 2017.

In an effort to distinguish itself from hundreds of other networks and broaden its appeal beyond the "family," ABC Family abandoned its name and branding strategy to become Freeform. In doing so, the network moved from targeting millennials to focusing on Becomers, an audience the network defines as broader in age and more media-literate. Whereas once live ratings were sufficient to determine whether a series was renewed on ABC Family, the dominant cultural capital in the digital, Peak TV era is a combination of ratings from live and flexible digital sources such as network apps and VOD, as well as social TV strategies expounded upon in Cassella, Galope, Brozek, and Proulx and Shepatin.

As ABC Family transitioned to Freeform, its senior executives repeatedly discussed the value and importance of having active viewers who engage with the series online. *Pretty Little Liars* provided the template, and *Shadowhunters* emulates its success. By focusing on organic and owned content, driving audience interest with hashtags, polls, and live tweeting, and capitalizing on the popularity of its stars' social media presence, *Shadowhunters* has managed to generate buzz and keep audiences engaged before, during, and after new episodes air, thereby ensuring its continued existence in a competitive television landscape cluttered with more than 450 television series. The alternative, as TV reporters were prone to say of *Chloe King*'s 2011 demise,[72] is to run out of lives.

NOTES

1. Cynthia Littleton, "Peak TV: Surge From Streaming Services, Cable Pushes 2015 Scripted Series Tally to 409," *Variety*, December 16, 2015, http://variety.com/2015/tv/news/peak-tv-409-original-series-streaming-cable-1201663212/.

2. Alan Sepinwall, "'Peak TV in America': Is There Really Too Much Good Scripted Television *Uproxx: What's Alan Watching*, August 8, 2015, http://uproxx.com/sepinwall/peak-tv–in-america-is-there-really-too-much-good-scripted-television/; Linda Holmes, "Television 2015: Is There Really Too Much TV?" *NPR*, August 16, 2015, http://www.npr.org/sections/monkeysee/2015/08/16/432458841/television-2015-is-there-really-too-much-tv.

3. Littleton, "Peak TV."

4. Maureen Ryan, "TV Peaks Again in 2016: Could It Hit 500 Shows in 2017?" *Variety*, December 21, 2016, http://variety.com/2016/tv/news/peak-tv-2016-scripted-tv-programs-1201944237/.

5. Carl DiOrio, "Fox Family Costs Mouse Less Cheese in Final Deal," *Variety*, October 24, 2001, http://variety.com/2001/tv/news/fox-family-costs-mouse-less-cheese-in-final-deal-1117854788/.

6. Ruth Suehle, "6 TV Networks That Aren't What They Started Out As," *Wired*, October 10, 2012, https://www.wired.com/2012/10/six-networks-that-changed/.

7. Caryn Murphy, "Secrets and Lies: Gender and Generation in the ABC Family Brand," in *The Millennials on Film and Television : Essays on the Politics of Popular Culture*, ed. Betty Kaklamanidou and Margaret Tally (Jefferson, NC: McFarland, 2014), 19–20.

8. *Ibid.*

9. Julie Liesse, "Getting to Know the Millennials," *Advertising Age*, July 9, 2007, http://brandedcontent.adage.com/pdf/ABC_Family_Meet_The_Millennials.pdf.

10. Michael Curtin, quoted in Murphy, "Secrets and Lies," 16.

11. Murphy, "Secrets and Lies," 22.

12. Sarah Fox, "Pretty Little Liars, Shadowhunters Make ABC Family Number One in Social Media," *The Slanted*, December 30, 2015, https://theslanted.com/2015/12/21909/pretty-little-liars-shadowhunters-make-abc-family-number-one-social-media/.

13. Elizabeth Wagmeister, "ABC Family to Rebrand Network 'Freeform' in January," *Variety*, October 6, 2015, http://variety.com/2015/tv/news/abc-family-freeform-rebranding-network-1201610697/.

14. *Ibid.*

15. *Ibid.*
16. Cynthia Littleton, "Upfronts: ABC Family Turns Up the Volume to Change Direction," *Variety*, April 14, 2015, http://variety.com/2015/tv/news/abc-family-pretty-little-liars-upfronts-1201472636/.
17. Wagmeister, "ABC Family to Rebrand."
18. *Ibid.*
19. *Ibid.*
20. Liam Burke, *The Comic Book Film Adaptation: Exploring Modern Hollywood's Leading Genre* (Jackson: University Press of Mississippi, 2015), 52.
21. For more on the genre of Teen TV, see Glyn Davis and Kay Dickinson, eds., *Teen TV: Genre, Consumption and Identity* (London: BFI, 2004) and Sharon Marie Ross and Louisa Ellen Stein, eds., *Teen Television: Essays on Programming and Fandom* (Jefferson, NC: McFarland, 2008).
22. Semantic justifications of genres have been reduced for space purposes. For more detailed, evaluative analyses of television genres see Jason Mittell, *Genre and Television: From Cop Shows to Cartoons in American Culture* (London: Routledge, 2004) and Glen Creeber, ed., *The Television Genres Book* (London: BFI, 2001). For analyses of teen television genre, see Patrick Bingham, "*Pretty Little Liars*: Teen Mystery or Revealing Drama?" n *Networking Knowledge* 6.4 (2014): 95–106.
23. Murphy, "Secrets and Lies," 21–22.
24. TV by the Numbers, "'The Nine Lives of Chloe King' Premieres on ABC Family June 14 at 9p ET," *TV by the Numbers*, May 4, 2011, http://tvbythenumbers.zap2it.com/network-press-releases/the-nine-lives-of-chloe-king-premieres-on-abc-family-june-14-at-9p-et/.
25. Nellie Andreeva, "ABC Family Greenlights 3 New Series—One Comedy and Two Dramas," *Deadline*, January 31, 2011, http://deadline.com/2011/01/abc-family-greenlights-3-new-series-one-comedy-and-two-dramas-102168/.
26. TV by the Numbers, "Tuesday Cable Ratings: 'Deadliest Catch' Tops; Premieres of 'Pretty Little Liars,' 'Memphis Beat,' 'Hawthorne,' 'Gene Simmons' & Lots More," *TV by the Numbers*, June 15, 2011, http://tvbythenumbers.zap2it.com/sdsdskdh279882992z1/tuesday-cable-ratings-deadliest-catch-tops-premieres-of-pretty-little-liars-memphis-beat-hawthorne-gene-simmons-lots-more/; Nellie Andreeva, "Cable Ratings: Soft Debut for 'Nine Lives'; 'Hawthorne' & 'Memphis Beat' Return Lower," *Deadline*, June 16, 2011, http://deadline.com/2011/06/cable-ratings-soft-debut-for-nine-lives-hawthorne-memphis-beat-return-lower-140545/.
27. TV by the Numbers, "Tuesday Cable: 'Teen Mom' Rises, Wins Again; TNT's Originals Rise; 'Pretty Little Liars' Dips," *TV by the Numbers*, August 17, 2011, 2016, http://tvbythenumbers.zap2it.com/sdsdskdh279882992z1/cable-ratings-teen-mom-up-wins-another-tuesday-absent-usa-originals-tnts-originals-rise-pretty-little-liars-dips/.
28. Nellie Andreeva, "ABC Family Cancels 'Nine Lives,' Picks Up 12 More Episodes of 'Lying Game,'" *Deadline*, September 15, 2011, http://deadline.com/2011/09/abc-family-cancels-nine-lives-171994/
29. Cassandra Clare's Shadowhunters, "About the Author," http://shadowhunters.com/about-the-author/.
30. Steve Baron, "ABC Family Orders Scripted Series 'Shadowhunters' Based on 'The Mortal Instruments' Series," *TV by the Numbers*, March 30, 2015, http://tvbythenumbers.zap2it.com/network-press-releases/abc-family-orders-scripted-series-shadowhunters-based-on-the-mortal-instruments-series/.
31. Ray Subers, "Weekend Report: 'Butler' Repeats, Newcomers All Open Below $10 Million," BoxOfficeMojo.com, August 25, 2013, http://www.boxofficemojo.com/news/?id=3718&p=.htm.
32. Ray Subers, "August Preview (cont.): 'Mortal Instruments,' 'World's End' & More," BoxOfficeMojo.com, July 31, 2013, http://www.boxofficemojo.com/news/?id=3708&p=.htm.
33. Nick Harley and Chris Longo, "44 Movies Being Adapted For TV," *Den of Geek*, January 18, 2017, http://www.denofgeek.com/us/movies/240179/44-movies-being-adapted-for-tv.

34. Deadline Team, "'Mortal Instruments' TV Series a Go with Ed Decter as Showrunner," *Deadline*, October 20, 2014, http://deadline.com/2014/10/mortal-instruments-tv-series-ed-decter-showrunner-855645/.

35. Patrick Hipes, "'Mortal Instruments' Series 'Shadowhunters' Lands at ABC Family," *Deadline*, March 30, 2015, http://deadline.com/2015/03/shadowhunters-tv-series-abc-family-mortal-instruments-1201401266/.

36. *Ibid.*

37. Ross A. Lincoln, "'Shadowhunters' Cast & Producers Pledge Faithfulness to Popular Novels—TCA," *Deadline*, January 9, 2016, http://deadline.com/2016/01/shadowhunters-mortal-instruments-freeform-1201679129/.

38. Dominic Patten, "'Pretty Little Liars' Ratings Rise Over 2015; 'Shadowhunters' Debuts Solid," *Deadline*, Jan 13, 2016, http://deadline.com/2016/01/pretty-little-liars-ratings-rise-shadowhunters-debuts-freeform-1201682580/.

39. Philip M. Napoli, *Audience Evolution: New Technologies and the Transformation of Media Audiences* (New York: Columbia University Press, 2010), 56.

40. Casey, "How Can TV Networks Maximize the Value of Social Media?"

41. Simon Dumenco, "Wait, Who's Actually Making Money Off Social TV? Beyond Metrics: How Networks Are Getting Marketers to Pony Up for Viewer Engagement," *Advertising Age*, September 17, 2012, http://adage.com/article/media/wait-making-money-social-tv/237223/.

42. Henry Jenkins, "Transmedia Storytelling And Entertainment: An Annotated Syllabus," *Continuum: Journal of Media & Cultural Studies* 24, no. 6 (2010): 944.

43. Ariane Galope, "#PrettyLittleLiars: ABC Family in TV's Post-Network Era" (MA dissertation, Victoria University of Wellington, 2016): 106.

44. *Ibid.*; Henry Jenkins, "'Do You Enjoy Making the Rest of Us Feel Stupid?' Alt.tv.twinpeaks, the Trickster Author and Viewer Mastery," in *Full of Secrets: Critical Approaches to Twin Peaks*, ed. David Lavery (Detroit: Wayne State University Press, 1995), Mary Kirby-Diaz, *Buffy and Angel Conquer the Internet: Essays on Online Fandom* (Jefferson, NC: McFarland, 2009).

45. Ryan Cassella, "The New Network: How Social Media is Changing—and Saving—Television," in *Television, Social Media, and Fan Culture*, ed. Alison F. Slade, Amber J. Narro and Dedria Givens-Carroll (Lanham, MD: Lexington Books, 2015), 2.

46. Napoli, *Audience Evolution: New Technologies and the Transformation of Media Audiences*, 72–73.

47. *Ibid.*

48. Lesley Goldberg, "Freeform's Fraught Year: Inside the Rebranded Network's Chase for Millennials (and a Hit Show)," *The Hollywood Reporter*, November 11, 2016, http://www.hollywoodreporter.com/live-feed/freeforms-fraught-year-inside-rebranded-networks-chase-millennials-a-hit-show-945184.

49. Sean Casey, "How Can TV Networks Maximize the Value of Social Media?" *Nielsen*, September 16, 2016, http://sites.nielsen.com/newscenter/how-can-tv-networks-maximize-the-value-of-social-media/.

50. Cassella, "The New Network," 3.

51. *Ibid.*, 4.

52. Stephen Harrington, Tim Highfield, and Axel Bruns, "More Than a Backchannel: Twitter and Television," *Participations: Journal of Audience & Reception Studies* 10, no. 1 (May 2013).

53. Mike Proulx and Stacey Shepatin, *Social TV: How Marketers Can Reach and Engage Audiences by Connecting Television to the Web, Social Media, and Mobile* (Hoboken: John Wiley & Sons, 2012), 19.

54. Sean Casey, "How Can TV Networks Maximize the Value of Social Media?"

55. Cassella, "The New Network," 9.

56. *Ibid.*

57. Todd Spangler, "The Most Social TV Shows of 2016," *Variety*, December 20, 2016, http://variety.com/2016/digital/news/2016-most-social-tv-shows-1201946167/.

58. Pretty Little Liars (@PLLTVSeries), Twitter, https://twitter.com/PLLTVSeries; Pretty Little Liars (@prettylittleliars), Facebook, https://www.facebook.com/prettylittleliars/.

59. Cassella, "The New Network," 7–8.

60. *Ibid.*

61. Melanie Brozek, "#PrettyLittleLiars: How Hashtags Drive the Social TV Phenomenon" (Senior thesis, Salve Regina University, 2013): 15–17.

62. Proulx and Shepatin, *Social TV,* 20; Cassella, "The New Network," 8.

63. Proulx and Shepatin, *Social TV,* 17.

64. Anjali Midha, "Study: Live-Tweeting Lifts Tweet Volume, Builds a Social Audience for Your Show," *The Twitter Media Blog,* September 18, 2014, https://blog.twitter.com/2014/study-live-tweeting-lifts-tweet-volume-builds-a-social-audience-for-your-show.

65. Fox, "Pretty Little Liars, Shadowhunters Make ABC Family Number One in Social Media."

66. Variety Staff, "'Shadowhunters' Renewed for Season 2 at Freeform," *Variety,* March 14, 2016, http://variety.com/2016/tv/news/shadowhunters-renewed-season-2-mortal-instruments-freeform-1201729260/.

67. TVSeriesFinale, "Freeform TV Show Ratings (updated 1/24/17)," *TVSeriesFinale,* January 24, 2017, http://tvseriesfinale.com/tv-show/freeform-tv-show-ratings-33382/.

68. Freeform—Shadowhunters Website, "Shadowhunters Season 2! Freeform Picks Up Shadowhunters for a Second Season!" March 14, 2016, http://freeform.go.com/shows/shadowhunters/news/shadowhunters-season-2-freeform-picks-up-shadowhunters-for-a-second-season.

69. Spangler, "The Most Social TV."

70. Ruth Kinane, "*Shadowhunters* Renewed for Season 3," *Entertainment Weekly,* April 21, 2017, http://ew.com/tv/2017/04/21/shadowhunters-renewed-season-3/.

71. Internet Archive Wayback Machine, "The Nine Lives of Chloe King Facebook Page," May 21, 2011, https://web.archive.org/web/20110521134411/http://www.facebook.com/TheNineLivesofChloeKing.

72. Kelly West, "ABC Family Cancels The Nine Lives of Chloe King," *Cinemablend,* 2011, http://www.cinemablend.com/television/ABC-Family-Cancels-Nine-Lives-Chloe-King-35133.html.

"Someone was watching us"
Surveillance and Spectatorship *in* Pretty Little Liars

Cara Dickason

The season one finale of *Pretty Little Liars* (2010–2017), "For Whom the Bell Tolls," opens with a tracking shot of a window, quickly identified as opening into the bedroom of Spencer Hastings (Troian Bellisario) when the camera glides toward the glass.[1] The titular Liars—Spencer, Aria Montgomery (Lucy Hale), Hanna Marin (Ashley Benson), and Emily Fields (Shay Mitchell)—are perfectly framed by the window pane, sitting atop Spencer's bed and crowded around a laptop. A cut moves the viewer into the room, closer to the girls, and another takes us over their shoulders to see what they are watching. On the laptop screen appears surveillance footage of their peer and rival Jenna Marshall (Tammin Sursok) blackmailing her stepbrother Toby Cavanaugh (Keegan Allen) into continuing their sexual relationship. This quick succession of shots takes up no more than the first fifteen seconds of the episode, but it encapsulates much of the tension at the heart of a series that centralizes narratives of surveillance, voyeurism, and visibility. *Pretty Little Liars* depends on the construction of girls as both subjects and objects of the gaze; girls look and are looked at in various configurations of power. Moments such as the one described above formally and narratively enact a negotiation between the distance of voyeurism and the intimacy of identification, between objectification and spectatorship. These categories are never entirely separate for the Liars or for the girl viewers who watch them, and *Pretty Little Liars* engages such spectatorial tension, both internally and externally to the diegesis.

Pretty Little Liars depicts a world in which its four young female protagonists are constantly subject to ubiquitous surveillance. Like two of its recent predecessors in teen girl television, *Veronica Mars* (2004–2007, UPN/CW) and *Gossip Girl* (2007–2012, CW), Freeform's most popular series addresses the dangers posed to girls by technology, as well as the powers it may afford. This essay aims to demonstrate how the integration of surveillance into the very fabric of the series' narrative and formal structures produces a distinctly feminine mode of address. By focusing on moments of non-consensual visual surveillance, I will suggest that the series' movement between points of view and between positions of visual power undermines traditional masculine scopic regimes. Its complex negotiation of subject and object positionality for the girls within the series seems to imagine a normative girl spectator whose active gaze must continuously confront her own objectification by the mainstream media. *Pretty Little Liars*, then, self-reflexively provides a fruitful ground to consider how girl viewers themselves negotiate spectatorship, surveillance, and representation.

The series extends its attention to girls' complex relationships to surveillance technology beyond the narrative to its transmedia marketing campaigns, though perhaps with less transparent reflexivity. The final section of this essay will argue that, despite the series' preoccupation with girls' dangerous exposure, *Pretty Little Liars* ultimately depends on its largely young female viewers to expose themselves to surveillance through social media activity and online consumption. In its own commercial interests, the series sells digital self-revelation, in contrast to the non-consensual visual surveillance in the show's narrative, as innocent, eschewing the similarly-objectifying effects of corporate data surveillance. Ultimately, *Pretty Little Liars* both internally and externally reproduces girls' active negotiation of viewership and visibility, shedding light on the complex relationships between girlhood, spectatorship, and surveillance.

The language of voyeurism and identification that I deploy here has its theoretical heritage in the psychoanalytic approaches to spectatorship of early feminist film theory. Such work has been thoroughly revised in recent decades, both to account for viewer agency and difference beyond a totalizing and deterministic gender binary and to consider how television and other media are structured by forms of pleasure other than the male gaze. Significantly, feminist television scholars Charlotte Brunsdon, Julie D'Acci, and Lynn Spigel suggest the possibility that TV historically constructs an essentially different subject position for the viewer than classical Hollywood cinema because "television as a medium is organized around female desire."[2] Its placement in the traditionally-feminized domestic space and long history of female-oriented genres such as the soap opera shifted television away from "soliciting visual pleasures that [objectify] female characters and [encourage] ways of seeing based on voyeurism and fetishism."[3] From this perspective, *Pretty Little Liars* typifies female-oriented teen television aimed at satisfying the viewing desires of an imagined normative, heterosexual female audience.

What makes *Pretty Little Liars* a particularly compelling object of study in this regard, however, is its conscious and active engagement with voyeurism as both primary mode of seeing and dominant televisual language through its incorporation of surveillance. By using voyeurism not only as a driving plot device but a formal technique, the series reflexively considers how young women navigate the kind of masculine scopic regimes first mapped by Laura Mulvey in her consideration of the objectifying male gaze in Hollywood cinema.[4] Not only do such unequally-gendered looking relations continue to dominate much narrative media, including television, they also operate in the larger systems of state and corporate surveillance ubiquitously deployed against girls and women.

Whether in film or television studies, however, theories that generalize about women's relationship to the gaze or appeals to feminine desire tend to collapse difference between and among women. Feminist spectatorship theory has thus increasingly tried to account for the agency of female viewers, including by addressing the resistant viewing practices of women of color and queer women—those outside the normative femininity most often represented and addressed by mainstream media (including *Pretty Little Liars*). Here, I employ Christine Gledhill's concept of negotiation as a "model of meaning production" as particularly salient for a series that thematizes girls' conflicted relationship to the gaze.[5] Negotiation contrasts the psychoanalytic theories of visual pleasure, which obscure differences within women's spectatorship and privilege white heteronormativity without naming it as such. As an analytic tool, negotiation creates space for the consideration of the complex spectatorial experiences of girls who do not directly identify with the normative white girlhood centralized in mainstream television; such girls experience the tension between visibility and voyeurism in ways both similar to and distinct from the girls on-screen. Gledhill proposes that "the term 'negotiation' implies the holding together of opposite sides in an ongoing process of give-and-take.... Meaning is neither imposed, nor passively imbibed, but arises out of a struggle or negotiation between competing frames of reference, motivation and experience."[6] The critical space of negotiation, inhabited both by the show itself and its young female viewers, allows for consideration of the role of surveillance in reflecting and constructing girls' ambivalent relationship to visibility, vision, and pleasure.

In the episode "For Whom the Bell Tolls," *Pretty Little Liars* formalizes the "competing frames of reference" to which Gledhill refers when the camera shifts from peering in through the window to offering the Liars' point of view, embracing their own voyeuristic gaze; in this scene and many others throughout the series, the show negotiates between girls as subjects and objects of the gaze, agents and victims of voyeurism.[7] The episode introduces the viewer to a series of mysterious surveillance videos, the ramifications of which reverberate throughout the series. At the end of the previous episode, "Monsters in the End," the Liars discover a small flash drive contained in a lunch box sitting at the center of a large, empty storage unit.[8] The stark, high-angle image from the girls' point of view highlights the irony of the immense power contained in such a small device. The flash drive physically represents the hyper-mobility of data (in this case, video surveillance files), its ability to be endlessly duplicated and to belong to anyone. In the scene that spans the end of "Monsters in the End" and the beginning of "For Whom the Bell Tolls," the Liars discover that the flash drive contains footage of them, obtained without their knowledge over many years. In Spencer's bedroom, a point-of-view shot of the laptop screen reveals a brief clip filmed from outside a window (recalling the opening shot of the Liars watching the laptop) of the girls playfully interacting in one of their bedrooms.

A reaction shot shows the girls' disgusted but hypnotized faces, solidifying the viewer's formal identification with their perspective in this moment, including their distressed reaction:

> ARIA: We're young girls in our bedrooms changing clothes.
> SPENCER: Exposed.
> HANNA: Do you think someone was watching us and getting off on it?
> SPENCER: Well, we all know who had a thing for younger girls.
> EMILY: I feel sick.[9]

Our identification with the Liars in this scene comes from our position as subjects of a shared televisual gaze (through the point-of-view shot) and extends to their position as objects of the gaze. The show here critiques the visual objectification of girls as well as the voyeuristic gaze that sexualizes them against their will. The show also refuses to show the most egregiously exploitative footage of the Liars; we do not see them changing clothes or more physically exposed than usual. We understand the invasion of privacy that occurred, but do not partake in the voyeuristic pleasure or discomfort of seeing them overtly sexualized. This may be in part due to the show's family-friendly network of ABC Family (as it was still titled when the episode aired), but is also intimately related to the show's target female audience.

Feminist film theorist Mary Ann Doane argues that female spectators cannot experience voyeurism in the way that male spectators do because "for the female spectator, there is a certain overpresence of the image—she *is* the image."[10] Within the diegesis, this overpresence is explicit; they are actually watching themselves through the lens of the voyeur and thus cannot experience any kind of pleasure from the image. But the show's formal strategy of showing us the video from their point of view, as well as the unwillingness to sexualize the Liars overtly, indicates an acknowledgment that this foreclosed experience of voyeurism extends to the show's female viewers. Our identification with the Liars watching the video supports our identification with the Liars exposed by the video, and vice versa. *Pretty Little Liars* thus enacts, through this moment of non-consensual visual surveillance, the spectatorial bind in which girl viewers are placed by mainstream media that ubiquitously displays their image and constructs them as objects for consumption.

The show emphasizes the relationship between ubiquitous visual surveillance of girls and girls' representation in the media, as well as the control that both forms of exposure exert. Girls' media scholarship and the work of Sarah Projansky, in particular, provide a lens through which to consider the distinct pleasures and anxieties that looking relations create for girls as opposed to women, although the line between the two categories is ambiguous in a culture that traditionally constructs girls as sexual objects and infantilizes women. Projansky uses the term spectacularization to describe how "media incessantly look at and invite us to look at girls. Girls are objects at which we gaze, whether we want to or not. They are everywhere in our mediascapes. As such media turn girls into spectacles—*visual objects on display.*"[11] Significantly, Projansky notes that "not all girls are spectacularized in the same way," suggesting that girls' relationships to visibility are largely determined by their race, class, sexuality, and other factors.[12] *Pretty Little Liars* primarily (though not solely) engages with a normative model of girlhood that privileges white, middle-class identity, and so it narrativizes the experience of those girls who perhaps identify more easily with the images most often spectacularized in mainstream media.[13]

Pretty Little Liars formally constructs the experience of confronting one's own incessant spectacularization in its depictions of the various headquarters of the Liars' anonymous antagonist, A. In the season two finale, "UnmAsked," Spencer enters a seedy motel room, the walls of which are plastered with photos of the Liars and Alison (Sasha Pieterse), the Liars' "Queen Bee" best friend whose disappearance and ostensible murder begins the series, as well as newspaper clippings about Alison's murder.[14] Sinister music climaxes as the camera circles Spencer, who stands in the center of the room and turns to see their pictures on all sides. The camera cuts to her point of view, panning in a circle to take in the overwhelming array of images; jump cuts within the pan, presumably from one wall to the next, create the sense that the pictures go on forever—there is no way to take them all in with a single shot. The scene visually and aurally heightens the sense of danger associated with surveillance as well as the feeling that these walls of images are closing in on Spencer. In this moment, surveillance and the voyeurism it enables pose an overwhelming threat to girls. The room also contains a dollhouse with five dolls resembling Alison and the Liars inside. The doll motif recurs throughout the series, culminating in the dollhouse storyline of seasons five and six in which A literally traps the girls in a house and forces them to behave per their wishes. "UnmAsked" reveals that the girls are controlled by A as toys in their game; the Liars are literally objectified, or made into objects. By visually pairing the dolls with the surveillance photos, the scene emphasizes the connection between visual objectification and the control of girls' bodies.

A similar scene in the season four mid-season finale, "Now You See Me, Now You Don't," finds the four liars entering what they believe to be A's new, much higher-tech lair.[15] In its conflation of visual and non-visual surveillance, the scene implicates everyday technologies that impose data surveillance in the dangerous objectification of girls. Blown-up photographs of Alison posing for the camera line the entranceway to the apartment, and inside, A has constructed large "timelines" for each of them. Spencer says, "All our secrets, private moments. A is documenting everything," as a series of close-ups reveal surveillance photos of the girls and their families all organized and labeled chronologically. The timelines recall Facebook's timelines, which are similarly meant to track important events through photos and posts in addition to being a social hub. The scene creates a connection, if not a direct comparison, between the sinister surveillance to which the Liars are constantly subject and the more mundane surveillance to which girls regularly subject themselves. However, as I will argue below, such implications are largely tempered through the show's more benign representations of everyday technology as well as its extra-diegetic engagement with social media.

A three-monitor computer system forms the centerpiece of the room, and Spencer tells the others, "That's how A's been watching us." Apparently discerning the contents of the computer screens, Emily adds questioningly, "A's monitoring the police? Watching the streets? Our alarm systems?" Spencer responds, "That's how A is everywhere." The combination of visual and data surveillance is of note here. As opposed to A's first lair, in which Spencer and the viewer along with her were bombarded by the sheer proliferation of their own images, this lair emphasizes the way that visibility is imposed through digital technology. There is no clear distinction made between A "watching the streets," perhaps by accessing CCTV feeds, and monitoring alarm systems. Both visual and data surveillance contribute to the ability to track, and potentially control, girls' movements, and thus their bodies.

A series of ensuing plot twists reveal that the apartment and surveillance equipment belong not to A, but to Ezra Fitz (Ian Harding), the teacher with whom Aria has an ongoing affair, who has secretly been investigating the Liars for a book he plans to write about Alison's murder. While this revelation makes little narrative sense, it does emphasize that the girls are subject to surveillance on multiple fronts. By the nature of an ongoing soap opera such as this one, the girls continue to seek out A and never truly uncover their identity. Even when the A they have long been searching for is finally revealed to be Alison's estranged, transgender sister (a development whose significant—and significantly underdeveloped—implications lie outside the scope of the present essay), she is quickly superseded by a new, more dangerous Uber A. Thematically, such delayed gratification leads to the sense that surveillance is truly inescapable, in part because it cannot be confined to a single source. Even as Spencer combs through the pictures in A's motel room, or Hanna tries to hack into Ezra's seemingly all-powerful computer, the Liars exist in a constant state of fear, anger, and disgust at their own objectification. They can resist the voyeuristic gaze of surveillance, but they cannot escape it. By formally and narratively identifying the viewer with the Liars in such moments, the show constructs a young female spectator deeply familiar with the experience of confronting her own spectacularization, imposed from every direction.

Just as we are aligned with the Liars' experience of objectification, however, the series continuously identifies the viewer with their active gazes. Even as *Pretty Little Liars* engages girls' negative reactions to voyeurism, it does not entirely eschew the pleasure that can be found in voyeurism as well. Returning once again to the opening scene of "For Whom the Bell Tolls," the camera takes on the Liars' point of view as it focuses entirely on the laptop screen showing the video of Jenna seducing and blackmailing her stepbrother Toby. A series of shot-reverse-shots show the Liars with eyes glued to the screen and the video in which Jenna sensuously removes Toby's shirt. The video ends after the viewer is given a long glimpse of Toby's bare chest, when Spencer disgustedly commands, "Turn it off." The Liars are certainly not meant to be understood as explicitly experiencing pleasure from watching this video, and again, the scene evokes what Doane calls "overpresence"; as fellow victims to surveillance, the Liars identify with Jenna's and Toby's objectification by the video and so do not take overt pleasure in it.

However, the scene momentarily revels in the power afforded to the voyeur. The Liars quickly turn to discussing how they will use the video against Jenna for their own benefit; like A, they subject Jenna to their voyeuristic gaze to manipulate her. Furthermore, even as the Liars do not exhibit any particular pleasure in their viewing, the lingering camera implicitly appeals to the assumed desire of the heterosexual female gaze, imposing on Toby a physical exposure the series rarely inflicts on its female stars. Significantly, we see Toby in a moment of vulnerability, while Jenna plays the dominant role, further emphasizing his status as object for consumption. In fact, the *Pretty Little Liars* camera frequently highlights the uncovered, well-sculpted male bodies that surround the Liars, acknowledging female heterosexual desire as a driving factor for girls within the show and those who watch it. Showrunner I. Marlene King has acknowledged the show's "wish fulfillment aspect" specifically with regard to the Liars' frequent relationships with older men, and the show actively acknowledges and works to fulfill girls' desire to look.[16]

In its embrace of female voyeurism, the series might seem simply to invert superficially the scopic regime of male-dominated narrative media without truly undermining the objectifying logic that empowers the traditionally-masculine position. I want to suggest, however, that the series' dynamic movement between such positionalities constitutes its feminine mode of address. Moreover, its attention to the desiring look of Emily, a lesbian, toward other young women in fact undermines the heteronormative paradigm of sexual difference on which traditional looking relations depend. Throughout the series, the camera often lingers on Emily's face as she looks at Alison, with whom she is infatuated, or one of her many romantic interests, identifying the viewer with her desire as much as with that of the heterosexual protagonists. Emily's girlfriends, however, do not receive the same kind of physically-objectifying treatment that the male love interests do, perhaps suggesting both an unwillingness to overtly sexualize lesbian desire and that lesbian desire offers an alternative to the unequal power dynamics implied by voyeurism in any configuration. In its various subversions and inversions of traditional voyeuristic relationships, *Pretty Little Liars* engages a distinctly feminine mode of looking.

Narratively, *Pretty Little Liars* works to identify the viewer with the four girl protagonists as both unwilling objects of the voyeuristic gaze of surveillance and subjects of their own desiring look. Formally, however, one of the show's most definitive conventions is its use of point of view shots of the Liars from afar, a strategy that aligns the viewer with the anonymous voyeurs constantly surveilling the girls. Our formal association with A's point of view is most explicit in the short, gimmicky scenes that end most episodes, in which the camera moves as if it were A's eyes. We see A's gloved hands extending from below, and the characters with whom they interact speak directly into the camera. Less explicitly, however, frequent tracking shots or long shots of the girls take on A's point of view or that of another antagonist, often cued by sinister music. Occasionally, these shots might appear to be taken on a (diegetic) recording device, with a flashing "record" marker in a lower corner, indicating the presence of formal surveillance technology. The subject of the point of view shot, if revealed by a reverse shot, may also be a character we suspect of foul play but who has not been confirmed as "for" or "against" the Liars, or the shot could serve to suggest a new suspect. Regardless, characters initially hidden by a point of view shot are not to be trusted, at least until proven innocent by one of the show's many elaborate twists. The series uses this formal convention to build suspense, as its central narrative mystery is the question of who is surveilling the Liars.

These point of view shots create a kind of exhilarating frustration for the viewer forced to take on an unknown perspective when we most identify with the characters being watched. To return to the opening shot of "For Whom the Bell Tolls," then, an interesting shift occurs when the tracking or long shot of the girls from outside a window or across the street is not revealed to be from anyone's point of view; the gaze simply belongs to the camera and to us, the television audience. The omniscient gaze of the series blends into the ubiquitous gaze of characters who are constantly surveilling the Liars.

Even as these conventions potentially highlight and exploit the voyeuristic gaze of the viewer, they may in fact serve to heighten a narrative paranoia that undermines the formal operation of voyeurism. Garrett Stewart suggests that the ubiquitous presence of surveillance technology in the film genre he dubs "surveillance cinema" actually serves to "deflect any twinge of voyeurism onto the diegetic paranoia of invaded privacy."[17] From this perspective, the scenes in which we take on A's (or another surveiller's) point of view allow us, the television viewers, to project our own voyeurism off of ourselves and onto A, the embodiment of the show's paranoia, even as we are implicated by the shared gaze. For Stewart, though, the deflection need not occur through such an explicit source of surveillance as A. He refers specifically to a scene in *The Bourne Legacy* (2012) in which the protagonist looks up at a hidden security camera someone has pointed out to him. Stewart writes, "No reverse shot locates the hidden camera. It's as ubiquitous as narrative omniscience itself."[18] Following Stewart's logic, the ambiguous relationship between A's point of view and the show's omniscient gaze allows the viewer's voyeurism to be subsumed into that of the surveilling characters in the narrative even when those characters are not diegetically present; we, the viewers, do not subject the Liars to voyeurism ourselves, but the surveillance that surrounds them does.

Stewart's theory certainly seems operative in *Pretty Little Liars*, although it does not account for the specificity of gender in voyeuristic looking relations. As mentioned above, Doane argues that female spectators cannot experience voyeurism in the traditional (masculine) sense. Writing about the "paranoid woman's film"—mainly classic melodramas such as *Rebecca* (1940)—she suggests that genres that centralize a female protagonist and place her in "a position of agency" inherently challenge her objectification and downplay the "to-be-looked-at-ness" of women in mainstream Hollywood film dominated by the male gaze.[19] Thus, Doane writes, "a certain despecularization takes place in these films, a deflection of scopophiliac energy in other directions, away from the female body … the aggressivity which is contained in the cinematic structuration of the look is released or, more accurately, transformed into narrativized paranoia."[20] To put Doane and Stewart in conversation, then, the frustration of the voyeuristic gaze in women's genres finds an outlet in narrative paranoia, which in turn allows the viewer to deflect their voyeuristic desires onto the diegetic surveillance technology. The presence of surveillance in *Pretty Little Liars* creates a cycle in which voyeurism is simultaneously condemned, engaged, and deflected. Such complex formal strategies eschew any straightforward embrace of the voyeuristic gaze that characterizes much media's traditionally masculine address.

In its overall embrace of the Liars' perspective, the series reflects on and critiques the danger posed to girls by ubiquitous surveillance and media representation. However, it does not deny the gaze of its female viewers even as the narrative paranoia, heightened by the show's formal conflation between the omniscient gaze of the camera and the voyeur's gaze, deflects any deep sense of voyeurism on the viewer's part. The integration of surveillance into the heart of the show's formal and narrative structure sheds light on girls' relationship to their own visibility and invites exploration of where there is room for girls' agency within regimes of constant visual surveillance. *Pretty Little Liars* actively negotiates, both formally and narratively, the tension between the complex positions of girls as subjects and objects of the gaze in a media landscape that insists on their spectacularization. The series thus engages a feminine mode of address to reflect and enact the struggle in which girls must engage to reconcile their own positions as subjects and spectators with a powerful cultural insistence on their objectification.

Not only does *Pretty Little Liars* construct its audience as spectators, however; the series understands girls as *users* of technology as well. The young female audience members of the series' target demographic have grown up in the internet era. They use Facebook and Twitter; many have smartphones that text, take pictures and video, and allow them to consume media and make purchases online. Much of their lives unfold in networked spaces that exist beyond traditional distinctions between private and public. In her essay "Surveilling the Girl via the Third and Networked Screen," Leslie Regan Shade notes, "the cell phone enables constant connectedness via texting and chatting, and has become an intrinsic facet in youth's everyday lives: its functionality fuses with identity formation and friendship safeguarding."[21] Digital connectedness, and the surveillance it entails, is an irrefutable fact of life for most girls—at least those middle-class girls targeted by mainstream media, including *Pretty Little Liars*. Technological visibility is inescapable and, for many, desirable, and the series engages this reality, responding to the way that their target demographic is understood to use technology. So even as *Pretty Little Liars* explores the potential of networked technology to objectify and control its young female protagonists, technology also functions positively, in mundane but significant ways.

Throughout the series, the Liars' relationship to personal technology is not always troubled by the threat of A's surveillance. For instance, Hanna complains to her mother, who refuses to buy her a new iPhone, "You're cutting me off from the world…. I can't go around without a phone. That's like going around without a brain. Or shoes." For Hanna (a serious shoe-lover), a phone is an extension of herself and the way that she connects to the world. Even though A seems to have access to the contents of all digital devices, cell phones and laptops on the show do not always carry the sinister undertones that assert the girls as objects of surveillance. Numerous extreme close-ups of digital screens not only serve to encourage the viewers' identification with the Liars' relationships to technology, but also to emphasize the centrality of technology to relationships and identity. Hanna, grieving over her break-up with her boyfriend Caleb (Tyler Blackburn), deletes him from her cell phone contact list, and we see her hand navigating the cell menu through her point of view. The dramatic moment signifies a decisive end to the relationship, and Hanna hesitates to go through with it. Cutting between Hanna's face and her view of the phone places the viewer squarely within the intimate relationship between a girl and the phone that exists as an extension of herself; deleting the contact is cutting Caleb out of her life.

Technology also serves to solidify their friendships and offer protection, particularly through the girls' frequent use of S.O.S. text messages. Whenever one of them sends an S.O.S., all of the other Liars flock to her no matter the circumstances. Aria reminds the girls to "keep your phones close" as they split up on a potentially dangerous mission in the season two finale, highlighting the security offered by their phones, as well as their necessity. Technology certainly serves as a site of exposure and constant surveillance on the show, but it simultaneously allows for intimacy, agency, and protection.

Not only does such a portrayal of the value of technology for girls respond to and reflect many girls' experiences, it aligns with the commercial needs of teen television series. This essay has so far addressed only the series' internal integration of surveillance and technology, but significantly, *Pretty Little Liars*' relationship to surveillance extends beyond the borders of the televisual text. As surveillance studies scholar David Lyon suggests, marketers and corporations behind TV shows "have a high level of interest in seeing surveillance sold as soft and benign."[22] *Pretty Little Liars*' transmedia marketing strategies rely on girls' access to technology and their willingness to make themselves vulnerable to surveillance through social media and other forms of online consumption and participation.

As many media scholars of teen TV suggest, the teen genre has a long history of transmedia marketing tactics. Referencing official paratexts such as exclusive interviews with the actors or affiliated music releases, Sharon Marie Ross and Louisa Ellen Stein assert that contemporary teen television is "essentially multi-media in nature."[23] According to Bill Osgerby, such strategies are part of a larger project of constructing and appealing to the most valuable consumers. Looking back to the advent of the teen girl genre,

Osgerby writes, "The torrent of 'teen girl' TV shows produced during the late 1950s and early 1960s, then, was just part of a wider business machine geared to reaping profit from a new, lucrative consumer market."[24] For this reason, early (and contemporary) teen TV shows featured white, middle-class characters; "teenage culture" was essentially the culture of a "leisure class," because that was the class seen as possessing the financial power that appealed to TV advertisers.[25]

Osgerby suggests that the femininity constructed by early teen girl TV shows "was characterised by a kind of 'consumerist hedonism,' young women using the products and resources of commercial youth culture to carve out a space for self-expression and personal pleasure."[26] Teen girl television series today, with *Pretty Little Liars* preeminent among them, are similarly invested in constructing a femininity based in consumption, although now technology drives more of that consumption. In an increasingly-saturated media market, network and cable television shows depend on young female viewers not only to watch the show (whether live, time-shifted, or online), but to engage in social media campaigns, join online fan communities, and purchase products advertised on their official websites. Such online activities expose fans to corporate surveillance that is then used to market directly back to them, reinforcing a model of femininity legible through acts of commercial consumption.

Pretty Little Liars has been hugely successful in its use of social media activity to expand and mobilize its fan base, and such tactics implicate viewers in the same regimes of surveillance the series' narrative navigates so complexly. As discussed above, however, everyday technology within the narrative is largely depicted as a site of autonomy and connection, and this depiction extends to the show's transmedia marketing campaigns. All of the lead actresses have massively popular Twitter and Instagram accounts with which to interact with fans, encouraging viewers' own online participation and self-revelation. Celebrity-driven tactics rely on a presentation of authenticity and the assumption that the actresses themselves control their social media profile. Troian Bellisario's Tumblr, for instance, contains her artistic photography alongside selfies and candid behind-the-scenes photos from the *Pretty Little Liars* set, constructing a carefully curated but ostensibly authentic sense of the actress who plays Spencer. The show's marketing promotes fan investment in knowing these actresses through their online presence, reinforcing social media as an important site of identity construction and self-revelation over which one can maintain control. Such tactics assert that, as Lyon suggests, "self-revelation is innocent, even as customer relationship marketing seeks new ways of inducing people to self-reveal to them"; they obscure the fact that one's social media profile is deeply vulnerable to surveillance.[27]

An early Twitter campaign for the show captures the dual positioning of girls as watchers and watched that dominates the show's narrative and its relationship to fans. Philiana Ng writes for *The Hollywood Reporter*, "ABC Family execs noticed viewers posting photos of As from their lives and launched an 'A Is Everywhere' contest in 2011, with a set visit as a prize; 753,554 fans posted on the Facebook app, generating 2.32 million views and 4,975 submissions. A poster of the top 50 images was created."[28] The campaign highlights girls as constantly subject to surveillance in their own lives and offers them a chance to resist publicly, exposing their As as A has presumably been exposing them. The title also serves as a reminder of the ubiquity of surveillance—one may not be able to escape A, but one can undermine their power. Social media becomes the tool by which one resists A's power to expose, but it is simultaneously the means by which one exposes oneself to further surveillance.

The show's most successful campaign to date, #WorldWarA, built up to the season four mid-season finale in August 2013. The hashtag was featured on-air and created suspense for an upcoming major reveal. The episode "Now You See Me, Now You Don't," which seemingly revealed Ezra as A, drew 1.97 million tweets, a record for any scripted television series.[29] Ng reports, "Knowing viewers shared photos and videos of reactions to past reveals, the network compiled a digital mosaic of Ezra on Facebook after the show aired."[30] The mosaic itself, constructed of thousands of fan reaction shots, serves to valorize fan visibility to some degree. Sending in a picture of oneself allows one to join the fan community and be a part of the show. However, it simultaneously turns that picture—and thus, that fan—into a pixel, a piece of data, visually metaphorizing the most prevalent form of surveillance—that which reduces people to data.

Pretty Little Liars narratively reveals the ways that visual surveillance can be deployed to manipulate, confine, and control girls. Data surveillance has the potential for similarly detrimental effects on girls, although the series obscures such effects in its extra-diegetic fan engagement. Lyon acknowledges that the data collected through social media or other online activity "may seem trivial (shopping preferences, for example), but when combined with others may help build a (rather partial) profile…. It is a profile that, in many cases, simply suggests what *sort* of person is here. The category, not the character, is all-important."[31] Certainly, much work in contemporary surveillance studies fails to consider deeply the specificity of women's or girls' relationship to surveillance, and Lyon's off-handed reference to the seeming triviality of shopping preferences obscures the extent to which normative femininity is largely constructed around consumer choices and enforced through surveillance. The targeted advertising companies use to profile online consumers, Lyon suggests, "may influence desires in new ways, [but] may also shape actual behaviours of certain social groups as individuals are encouraged by feedback to fit the expected patterns."[32] Feminine social norms are continually reinscribed by the corporations and institutions that financially benefit from a consumption-based ideal of girlhood.

Zygmunt Bauman reiterates that surveillance-based marketing practices are a "restrictive, panopticon-style undertaking," but notes that "all that targeting, of course, only applies to fully fledged, fully feathered consumers."[33] Those determined by surveillance not to be of consumer value are excluded from the economy of exchange that does, in fact, confer privilege onto those who embody the proper forms of consumption. The "data-double" constructed out of a person's online or networked choices and movements "refers to individuals but simultaneously is only a kind of tool—useful or not to institutions," and "those who don't conform to expectations" are cut off.[34] While capitalist society has long viewed white, heterosexual, middle-class and otherwise normative girls as a valuable consumer market, data surveillance serves to further marginalize girls outside of those categories and girls unwilling or unable to participate in ways "useful" to institutions.

Such dynamics are as central to girls' relationship to their own subjectivity and visibility as their confrontation with ubiquitous visual objectification. *Pretty Little Liars*' commercial interests, however, only deem the latter eligible for narrative fodder. Lyon suggests that "contemporary surveillance occurs in contexts that are already media-saturated…. Some [media] help us grapple with the 'gaze' more intelligently, some help the gaze drift out of focus. Some aid critique, some aid complacency."[35] *Pretty Little Liars* compellingly and self-consciously grapples with the gaze through its depiction of surveillance. Ultimately, it aids both critique and complacency, engaging girls' struggles to control the terms of their own visibility while simultaneously encouraging girls to expose themselves to consumer-driven surveillance that undermines their agency. Narratively, the series relies on its female protagonists to act and, in doing so, resist their object status. But commercially, it may in fact depend on some level of complacency from viewers willing to subject themselves to surveillance. However, it is reductive to view girls complicit in their own surveillance as "mere unwitting dupes of a capitalist conspiracy," and *Pretty Little Liars* certainly constructs a spectator resistant to her own objectification.[36] The series seems to invite the viewership and identification of girls who must constantly negotiate their own representation, or lack thereof, in the media they consume, and who take pleasure in their own gaze even as they approach the power dynamics of surveillance critically. *Pretty Little Liars* is inherently a commercial endeavor dependent on the digital surveillance of its viewers, but through its engagement of a feminine mode of address, it invites and constructs a critical, flexible, and active spectatorial gaze. The relationships between subjectivity and objectification, between spectatorship and surveillance remain in tension, and *Pretty Little*

Liars' girl viewers possess, though not without challenge, the agency to explore and negotiate that tension.

ACKNOWLEDGMENTS

Many, many thanks to Caetlin Benson-Allott, Lori Merish, Brian Hochman, and my Georgetown colleagues for their guidance in the initial writing of this work, and to Mimi White, Nick Davis, Lauren Herold, the Northwestern Gender and Sexuality Studies Graduate Colloquium, and the Screen Cultures program for their help in its extensive revision. Thanks as well to Suzanne Reed for her encouragement, and Diane Rizk and Dana Dickason for their support.

NOTES

1. "For Whom the Bell Tolls," *Pretty Little Liars*, season 1, episode 22, directed by Lesli Linka Glatter, written by I. Marlene King, ABC Family, 21 March 2011.

2. Charlotte Brunsdon, Julie D'Acci, and Lynn Spigel, *Feminist Television Criticism: A Reader* (Oxford: Oxford University Press, 1997), 6.

3. *Ibid.*

4. Laura Mulvey, "Visual Pleasure and Narrative Cinema," *Screen* 16, no. 3 (1975).

5. Christine Gledhill, "Pleasurable Negotiations," in *Cultural Theory and Popular Culture: A Reader*, 3d ed., ed. John Storey (Essex: Pearson Education Limited, 2006), 114.

6. *Ibid.*

7. "For Whom the Bell Tolls," *Pretty Little Liars*.

8. "Monsters in the End," *Pretty Little Liars*, season 1, episode 21, directed by Chris Grismer, written by Oliver Goldstick and Maya Goldsmith, ABC Family, 14 March 2011.

9. "For Whom the Bell Tolls," *Pretty Little Liars*.

10. Mary Ann Doane, *Femmes Fatales: Feminism, Film Theory, Psychoanalysis* (New York: Routledge, 1991), 22.

11. Sarah Projansky, *Spectacular Girls: Media Fascination and Celebrity Culture* (New York: New York University Press, 2014), 5.

12. *Ibid.*, 7.

13. *Pretty Little Liars* compellingly addresses non-normative girlhoods only in terms of Emily's homosexuality and its effect on her vulnerability to surveillance, largely obscuring any effects of racial or class differences (both of which Emily embodies as well). The role of race, class, and sexuality in determining girls' relationships to surveillance on the show warrants further consideration.

14. "UnmAsked," *Pretty Little Liars*, season 2, episode 12, directed by Norman Buckley, written by I. Marlene King, ABC Family, 19 March 2012.

15. "Now You See Me, Now You Don't," *Pretty Little Liars*, season 4, episode 12, directed by Lesli Linka Glatter, written by I. Marlene King and Bryan Holdman, ABC Family, 27 August 2013.

16. Jessica Goldstein, "Showrunner Marlene King on *Pretty Little Liars'* Confusing Weather and Its Many Older Men," *Vulture*, June 10, 2014, http://www.vulture.com/2014/06/showrunner-marlene-king-interview-pretty-little-liars-crazy-seasons-and-many-older-men.html.

17. Garrett Stewart, "Surveillance Cinema," *Film Quarterly* 66, no. 2 (Winter 2012): 11.

18. *Ibid.*

19. Doane, "The 'Woman's Film,'" in *Re-Vision: Essays in Feminist Film Criticism*, ed. Mary Ann Doane, Patricia Mellencamp, and Linda Williams (Los Angeles: The American Film Institute, 1984), 70.

20. *Ibid.*

21. Leslie Regan Shade, "Surveilling the Girl via the Third and Networked Screen," in *Mediated Girlhoods: New Explorations of Girls' Media Culture*, ed. Mary Celeste Kearney (New York: Peter Lang, 2011), 263.

22. David Lyon, *Surveillance Studies: An Overview* (Cambridge: Polity Press, 2007), 154.

23. Sharon Marie Ross and Louisa Ellen Stein, eds., *Teen Television: Essays on Programming and Fandom* (Jefferson, NC: McFarland, 2008), 12.

24. Bill Osgerby, "'So Who's Got Time for Adults!' Femininity, Consumption and the Development of Teen TV—from *Gidget* to *Buffy*," in *Teen TV: Genre, Consumption and Identity*, ed. Glyn Davis and Kay Dickinson (London: British Film Institute, 2004), 75.

25. *Ibid.*, 77.

26. *Ibid.*, 82.

27. Lyon, *Surveillance Studies*, 154.

28. Philiana Ng, "'Pretty Little Liars': Inside the Bold Strategy of Getting Teens to Watch TV," *Hollywood Reporter*, July 7, 2014, http://www.hollywoodreporter.com/news/pretty-little-liars-inside-bold-714481

29. Kat Ward, "Chart Attack," *Entertainment Weekly*, September 12, 2014, 19.

30. Ng, "'Pretty Little Liars.'"

31. Lyon, *Surveillance Studies*, 101.

32. *Ibid.*

33. Zygmunt Bauman and David Lyon, *Liquid Surveillance* (Cambridge: Polity Press, 2013), 126.

34. Lyon, *Surveillance Studies*, 114; Bauman and Lyon, *Liquid Surveillance*, 123.

35. Lyon, *Surveillance Studies*, 155.

36. *Ibid.*

Forever Young
Youth Fetishism in *the* Shadowhunters *Universe*

STEPHEN P. SMYTH

The opening sequences from the pilot episode of Freeform TV's *Shadowhunters: The Mortal Instruments* series (2016–present) present viewers with a looping narrative full of manifold representations of youth and youthfulness. Three young mystical warriors make their way acrobatically through a crowded city street. They pursue a shapeshifting demon who grows progressively younger, transforming from a middle-aged businessman into a twenty-something club vixen. Meanwhile, our heroine, Clary Fray (Katherine McNamara), is serenaded by her best friend Simon (Alberto Rosende) on the night of her eighteenth birthday with a rendition of Alphaville's 1984 single "Forever Young." These are but a handful of examples from the stream of youthful signifiers that suffuse the *Shadowhunters* narrative from the opening of the series onward.[1] The first season of *Shadowhunters*—featuring Clary on her journey from life as a "mundane" (normal human) into the magical realm of the Shadow World, where she discovers she is herself a half-human, half-angel Shadowhunter—mercilessly immerses the viewer in ineluctable currents of youthfulness, longevity, and rejuvenation, so much so that *youth* as a reified construct becomes a sort of mold within which the entire series is cast.

As such, youth, or the fetishization and longing for it, becomes the form and substance of the series, that with which the viewer engages as much or even more so than any storyline or morally relevant message. The end result is that viewers identify with and yearn for the youth of each character as well as the actors in their roles. In this way, youth, or the promise of engaging with it, renders *Shadowhunters* a factory for producing desire—for youth, longevity, and in the end, more desire itself—with the cast and characters acting as workers in an assembly line production process.

The series, which follows on the tail of the critically and commercially lackluster film *The Mortal Instruments: City of Bones*, is based on the first book of six in *The Mortal Instruments* young adult (YA) novel series by Cassandra Clare (pen name for author Judith Lewis née Rumelt).[2] Clare did not initially intend the series as YA genre fiction, but rather adult literature featuring younger characters. However, her publisher chose to place it in the YA category as a marketing tactic. While not intended as YA, *The Mortal Instruments* series nevertheless relies on many YA conventions, such as the struggles of a young female ingénue coming of age in an almost alien world, star-crossed romance, emerging sexuality, encroaching adult responsibility, and a generally pervasive youthfulness across characters and subject matter. The prior film adaptation was a nearly direct appropriation of the book material, and yet it failed to appeal to its intended audience. The TV series, on the other hand, uses Clare's books for inspiration while making many adjustments to character age and desirability, subject matter, staging, and key plot points. Such changes were made ostensibly to appeal to Freeform's target 14–34 age demographic while also compounding the youth fetishization inherent in the source material.

This essay explores the nature of such youth fetishization, as well as the relationships between the production and consumption of youth, both within the *Shadowhunters* universe and in the world of consumer capitalism beyond the series' fourth wall. By examining the development of youth consumption from the time of Karl Marx to the millennials, we will see that within a primarily consumer culture there is little need for adults at all. Indeed, throughout the first season of *Shadowhunters*, we witness this message time and again, as the older generation, with its stern, inflexible, and outdated moral imperatives effectively creates the chaos and adversity that the much more resilient, resourceful, and inventive younger Shadowhunters must continually put right. This attitude that privileges youth over age, the *up*dated over the perpetually *out*dated, creates a dialectical system that outmodes the just-recently-new with ever-increasing rapidity. This is true not only of the *Shadowhunters* narrative, but also the YA genre form which it hails and the world of consumer capitalism, which renders youth both desirable and consumable.

In the Shadow World[3]

In the *Shadowhunters* fictional universe, everyday reality is but a veil over a world in which all that we believe to be fantasy is in fact true. This deeper reality is known by its denizens as the Shadow World. As we learn during season one, a constant struggle ensues within the Shadow World between order and chaos, with the Shadowhunters, or *Nephilim*, of the Clave (the ruling bloc of Shadowhunter society) as self-proclaimed keepers of order and demons as agents of chaos. In between these two factions reside mundanes and Downworlders, such as faeries, warlocks, werewolves, and vampires, over whom the Clave reigns in the course of keeping that order.

Order in the Shadow World is maintained through the Accords, peaceful treaties between the Clave and the various Downworld races, arrived at after centuries of bloodshed on all sides. However, not all Shadowhunters agree with the Accords. Once such dissenter is Valentine Morgenstern (Alan van Sprang), a rogue Shadowhunter who seeks to eradicate not only all demons, but all Downworlders as well. Through dialogue and a series of narrative flashbacks, we learn that as a young man, Valentine became the archenemy of the Clave, and along with his followers, who identified as the Circle, advocated and eventually engaged in open rebellion against Clave leadership.

The series pilot episode opens on the day of Clary's 18th birthday, during which she discovers the Shadow World while embroiled in a skirmish between several demons and the three young Shadowhunters, Jace, Alec, and Isabelle (Dominic Sherwood, Matthew Daddario, and Emeraude Toubia, respectively) introduced in this essay's opening passages. Clary then learns that she was born a Shadowhunter, raised in the mundane world by her mother, Jocelyn (Maxim Roy), to protect her from the notorious Valentine, who, as it turns out, is Clary's father. During episode one, members of the Circle kidnap Jocelyn, who has drunk a warlock-concocted sleeping potion, and deliver her comatose body to Valentine, who then holds her prisoner.

Valentine seeks to discover the whereabouts of the sacred Mortal Cup, the vessel capable of creating new Shadowhunters from mundanes. Through yet more flashbacks we discover that Jocelyn absconded with the Cup over a decade previously in order to prevent Valentine from generating a rebel army of new Shadowhunters. Clary discovers Jocelyn hid the Cup and that Valentine and other Downworlders now believe Clary knows of its whereabouts. However, Clary, having no such knowledge, believes she can use the Cup as a bargaining chip for her mother's life, if only she can locate it first.

The remainder of the first season's narrative arc follows Clary on her quest for her mother and the elusive Mortal Cup, as well as on her path into emerging adulthood. It is at the point of this emergence that Clary's youthful naivety becomes most palpable as she struggles to reconcile her adolescent idealism with the realities of her newly imposed adult role. Further, the central narrative importance of the Mortal Cup, the vessel that creates new Shadowhunters, foregrounds the importance of newness and rejuvenation to the overall story, one in which all roads lead to the Cup and therefore to the source of new life. In this way, both the Cup and Clary's heroic journey become but the most salient of youthful symbols that permeate the *Shadowhunters* universe.

Youth Fetishism

In his book *Consuming Youth: Vampires, Cyborgs & the Culture of Consumption*, Rob Latham examines the popularity of vampire and cyborg imagery in contemporary literature, film, and other media, noting the ways in which this imagery contributes to and reflects a culture obsessed with fetishizing and literally consuming youth. By invoking Latham's writing, my goal is to establish a groundwork for understanding the inception and rise of youth culture within consumer capitalism and the contemporaneous increased popularity for young adult, urban science fiction, and fantasy fiction, such as *Shadowhunters*. While it has been hyperbolically suggested that *all* American fiction is young adult fiction,[4] *Shadowhunters* represents an extreme example of youth fetishism within the genre.

In defining youth fetishism, Rob Latham invokes the work of W. F. Haug, who coined the term to describe several key facets of commodity consumption in twentieth century capitalist cultures. For Latham and Haug, youth fetishism refers to "'(1) the compulsive character of the young'; (2) sensualized images of youth to provoke consumer appetite; and (3) a pervasive ideology of youthfulness, which 'subjects the whole world of useful things, in which people articulate their needs in the language of commercial products, to an incessant aesthetic revolution.'"[5] While these ideas permeate Haug's writing, it was Latham who tied them together into a coherent model for identifying and defining youth fetishism, a notion which was but one facet of what Haug characterized as the "moulding of sensuality" of commodities under capitalism.[6] Whereas Karl Marx wrote of exchange and use values of commodities, where use value describes that actual usefulness of a commodity to the buyer and exchange value indicates the price that can be gotten by the seller for said commodity,[7] Haug argues that in current capital exchanges, the so-called "promise of use value" is all that matters.[8] In other words, the commodity need have no real inherent use value, but rather the appearance of such through its association with some sensual quality or experience. One way to achieve this, according to Haug, is to accentuate the object's newness and to cast similar objects that came before it as outdated. Haug referred to this process as "aesthetic innovation" of the newer commodity and "aesthetic aging" of that commodity's older counterpart.[9] In this way, fetishizing the commodity's "youth" relative to other similar items makes the newer item more desirable.

Unpacking the evolution and components of youth fetishism, Latham explains that Marx first noted the connection between his theories of political economy and the vampire/cyborg dialectic in his writings on capitalism, albeit in a rather primitive manner. Marx used the image of the bloodsucking fiend of European folklore to symbolize "dead labor, which, vampire-like, lives only by sucking living labor," and the human-machine or cyborg trope to characterize a system in which "workers are merely conscious organs" within the industrial factory.[10] Latham explains how, for Marx, the vampire archetype derived more from the "darker, more brutal figure of central European folklore" than the romanticized, aristocratic version that arose after John Polidori's *The Vampyre,* itself a satire of the decadent life of Lord Byron.[11] This distinction becomes ever more important as the focus on capitalist culture shifts from production to consumption in twentieth century economic discourse, with an equally increasing emphasis on vampires in film and literature as figures of desire and seduction, rather than the folkloric, mindless bloodsuckers from Marx's analysis.[12]

In terms of cyborg imagery, the emergence in the twentieth century of cybernetic theory allowed for a reinterpretation of Marx's conflated worker-machine, casting Marx's model as a "protocybernetic system."[13] As Latham explains, capitalist production developed a series of crises "essentially of communication and control ... which necessitated automation measures that now seem less industrial than implicitly cybernetic."[14] The diametrical tension between the vampire and cyborg archetypes generated what Rob Latham describes as a "perfect dialectical image in which unprecedented technical progress and primitive, inhuman exploitation coexist in a structure of profound contradiction."[15]

Within the *Shadowhunters* narrative, we see the vampire-cyborg dialectic play out in interactions between the Shadowhunter and Downworld cultures. The Shadowhunter base of operations, the New York Institute, while appearing as a dilapidated Gothic Revival church to outsiders, houses a high-tech fortress inside, which is the domain of the young. One of many deviations in the adaptation from the *Mortal Instruments* book series to the *Shadowhunters* TV program is the conversion of the Institute from a low-tech, archaic Nephilim sanctuary, run only by the Lightwood family and Hodge (Jon Cor) in the novels, to the TV version of the Institute, featuring advanced lighting systems embedded in floors and ceilings, ultra-modern glass and steel interior partitioning, and state of the art surveillance equipment with myriad digital video touchscreen interfaces papering the walls and operated by numerous young workers. This alone endows the Shadowhunter workday with cybernetic superpowers.

Shadowhunters are further aided in their duties by the use of magical runes, the so-called "Marks of Raziel," derived from a "complex runic language" and given to them by the Angel Raziel, granting each Shadowhunter special powers beyond those of typical mundanes.[16] Each warrior burns these runes into their own skin in order to enhance physical and mental abilities with a wand-like *stele*, a tool made of a special "heavenly metal"[17] known as *adamas*.[18] There are runes for healing, for greater strength and agility, even a rune for seeing in the dark. While the runes etch the skin and resonate with archaic power, they augment the Shadowhunters' abilities as much as the bionic limbs of *The Six Million Dollar Man*. Runes make the Shadowhunters better, stronger, faster, even invisible. At the same time, as arcane relics from a bygone age which also give the young warriors the enhancements of cutting edge technology, the runes carry a similar theme to the vampire-cyborg in and of themselves, in that they create a dialectical tension between the very ancient and the ultramodern. More than any high-tech device or tattooed rune, however, the Shadowhunters share an almost robotic devotion to selfless duty, unquestioning loyalty to the Clave, and the suppression of human emotions and desires.

In contrast to the cyborg-like nature of the Shadowhunters, many Downworlders possess more vampiric traits. In the Shadow World, both vampires and werewolves are almost single-mindedly predatory, preternaturally strong and vital, and possessed of a certain blood lust. Vampires and warlocks live lavish, indulgent lifestyles, making them the superlative consumers and reapers of profit within the Shadow World. While warlocks provide services for pay, vampires seem to live off of their investments. As the vampire Raphael (David Castro) tells Simon in episode three, "let's just say we invest early and often."[19] Faeries are less opulent, though they live within lush surroundings, at least as far as we can tell from Meliorn (Jade Hassouné), the sole faerie we meet throughout the first season. In general, many Downworlders appear a bit shady, often harboring hidden and nefarious motives, which again is more emblematic of the vampire trope than that of the cyborg.

Therefore, while the Clave is primarily cybernetic, the Downworld is mostly vampiric. These two forces work to maintain a tense harmony in which neither group particularly trusts the other, with each laboring under the awareness that peaceful coexistence is better than the alternative. As such, they represent the tense-but-balanced vampire-cyborg from Latham's reading of Marx. In this system, the Clave maintains hegemony over the Downworld, not through financial incentive, but rather through the promise not to wipe them all out.

From Marx's nineteenth century capitalist-as-vampire and worker-as-machine dialectic, we move to a twentieth century, postwar Fordist culture, characterized by the two forces of "mass production and mass consumption,"[20] or what Jean Baudrillard described as "productivity" and "consumativity."[21] In this system, ruled, according to Baudrillard, by desire "abstracted and atomized into needs," production becomes a "by-product of its productivity." In other words, rather than being the driving force within the economic system, production is a necessary expedient to meet the demands for consumption. For Baudrillard, consumativity drove the production process, and the masses became enslaved by their need to consume, that is, to fulfill their desires.[22]

The term Fordism was coined by Antonio Gramsci in his 1934 essay "Americanism and Fordism," in which he described Fordism as a uniquely American system that dispensed with the "vicious parasitic sedimentations" of old world European upper class privilege and wealth, allowing an expansive middle class to grow and prosper.[23] David Harvey cites the "symbolic initiation date of Fordism" as 1914, when "Henry Ford introduced his five-dollar, eight-hour day for workers."[24] Under the Fordist regime, the combination of higher wages for laborers and reduced prices due to mass production allowed average consumers to afford more products, beyond what was needed for daily survival.[25]

In its nascent stages, around the turn of the twentieth century, the ideal Fordist producer possessed an adult work ethic that mandated know-how, punctuality, and reliability. Consumption, on the other hand, emphasized play and leisure, which encouraged a more childlike or youthful attitude. As a result, a contradiction arose in which workers were expected to be simultaneously adult producers and childlike consumers. Over time, the assembly line system demanded less and less specialized knowledge, leading to a diminished need for adult-like qualities in the workforce. In turn, an ethic emerged during the prewar period that demanded skills such as "quick reflexes, endurance, and adaptability, nominally associated with youth."[26] Thus, what began as tension between the responsibilities of work and the sybaritic indulgences of leisure, or between the adult producer and childlike consumer, trended toward attitudes that supported the construct of youth, both at work and in the domestic sphere.[27] The growing emphasis on youth under Fordism led to its inflated status as a desirable trait, which in turn engendered the rise in the fetishization of youth, both in work and leisure. In the postwar period, the "cyborgization of youth" arose as youthful imagery was conflated with production to an ever greater degree,[28] while the primitive qualities of the vampire became overshadowed by the voracious appetite that infected consumerism.

David Harvey argues that Fordism continued to dominate the economic landscape until about 1973, when what he referred to as a period of "flexible accumulation" took its place.[29] Others, such as cultural theorist Stuart Hall, refer to this period as post–Fordism.[30] According to Hall, post–Fordism consists in part in a shift to information technologies, a decline of the manufacturing base, growth of computer-based industries, an increased emphasis on targeting consumers by lifestyle, taste and culture, the rise of the service and white-collar classes, the feminization of the workforce, and globalization facilitated through the communications revolution.[31] Latham argues that as we shift ever further toward an information based society, particularly in the post–Fordist era, "contemporary experience becomes an ongoing acculturation into cyborg possibility," with youth remaining "on the front lines."[32]

The young Shadowhunters exemplify this notion of youth, fighting at the front lines of a fictionalized culture war, keeping the vampiric forces of the Shadow World in check through their cyborg-like technology and runic magic. Of course, these youthful themes permeate the Downworlders as well, each side containing within itself the trappings of youth and the accompanying fetishized power and desirability therein. Whereas the older generations seek to subdue and even eradicate the Downworlders, the younger Shadowhunters often forge alliances with their counterparts such as Luke (Isaiah Mustafa), who we learn is a werewolf, Simon, once he becomes a vampire, and Magnus Bane (Harry Shum, Jr.), the High Warlock of Brooklyn, among others. Thus the lines between Shadowhunter and Downworlder become less distinct. While this has its advantages in the fictional realm, in the world of post–Fordist capitalism, this continued merging of cybernetics and vampirism, that is, technology and exploitation, can have some very insidious side effects. Whereas Marx envisioned a world in which the cybernetic aspects of the factory could be mastered by the masses who would then overthrow the exploitative aspects of the vampire-capitalists, if the Shadow World is reflective of the real world in any sense, the cyborg and the vampire appear to remain symbiotically intertwined.

Youth Saturation

Even as the vampire-cyborg dialectic plays itself out in many direct and symbolic ways in the *Shadowhunters* narrative, the saturation of youth moves beyond these strict dialectical confines and spreads across multiple aspects of the narrative and production. Each major character struggles to leave childhood behind and to define what adulthood means to her or him, while dialogue and plot scenarios speak of a pervasive attitude that the older generation is misguided and youth knows better. Vampires, warlocks, and faeries are youthful and beautiful. The title characters are young, athletic, and fit. Magnus, whose age is variously cited at somewhere between three and four hundred years, appears as a man in his thirties, makes great strides to wear the most fashion-forward apparel, resides in a trendy New York loft, muses on the paradoxical nature of his appearance versus his actual age—"I was alive when the Dead Sea was just a lake that was feeling a little poorly"[33]— all while courting twenty-two-year-old Alec. Judging from Magnus, youth, for a warlock, is more a matter of appearance than chronology.

Various plots and subplots focus attention on innocence, coming of age, and the assumption of newly-imposed adult roles. Indeed, the primary narrative arc, which situates Clary as the heroine whose mythical journey begins at the moment of her coming into adulthood, is entirely about the struggle to find her own way in the world. For Clary, this process entails lifting the veil from the Shadow World, seeing all things as they truly are, and responding to the demands placed in front of her with maturity and clarity.

This journey from adolescence into adulthood is shared by others within the *Shadowhunters* narrative as well. Indeed, each of the show's primary characters struggles in their own unique ways throughout the course of the first season to figure out just what being an adult means. Jace grapples with his personal identity after discovering Valentine is his father and Clary his sister; Alec and Isabelle each seek balance between desire and duty while laboring under the panoptic gaze of their judgmental mother; and Simon agonizes over the divergent forces tearing at him from both inside and out after being turned into a vampire by Camille (Kaitlyn Leeb). These struggles foreground not only the characters' emergent adulthood, but also their palpable youth and naive inexperience. Ironically, it is this lack of adult savvy that becomes each character's greatest strength as it leads them to invent their own paths and to see the folly of their parents' beliefs and practices. As such, finding access to one's personal authenticity in the midst of moral and ethical ambiguity becomes each character's driving motivation.

At the core of Clary's struggle lies a constantly shifting, decentered sense of subjectivity and a moral compass that spins so fast it makes one dizzy. The mundane reality Clary thought she knew disappears, replaced by a confusing mirror world where very little makes sense to her. In Clary's new existence, her evil father Valentine claims the Clave is corrupt, that he is the moral authority, and that all Downworlders are condemned to iniquity via their tainted blood. The Clave vilifies Valentine for his racist extremism while simultaneously espousing elitist and bigoted ideas about Downworlders themselves. At the same time, many Downworlders *do* exhibit reprehensibly stereotypic behavior, though often out of fear and defensiveness. The Clave eventually conducts witch hunts and show trials in a ploy to wrest the Mortal Cup from Clary once she's found and hidden it from them, betraying their own deeply held values in their desperation to maintain power and order. Clary sees all this and constantly questions the lack of fairness, ethical consistency, and social justice present within the Clave's decision making. This leaves Clary grasping about for adult role models to guide her, but finding time and again that she must forge her own unique identity and rely upon her own values, judgments, and instincts in the process.

This shifting terrain affects all of the young characters who watch the elder generation lay down a clear, ideological framework and then flagrantly trounce all over it in the name of politics and family honor. Like Clary, Jace discovers his entire childhood was a lie and the brutal lessons taught by his father—whom he now knows to be have been Valentine—were based on lunatic fanaticism. Similarly, when Alec and Isabelle discover their parents to be former Circle members, they find themselves questioning all of their previous assumptions about duty and honor. For Alec, this pushes him into defiance against the establishment he had faithfully served, whereas for Isabelle, it drives her to emulate her mother, who, like Isabelle, was once motivated more by passion than reason. The primary characters, and thus the audience who identifies with them, grasp at that shifting moral center for some sort of truth or sense of ethical direction, but being continually thwarted, are left ever more confused about the right and wrong of it all. This sense of confusion, of grasping for a sound ideology or rootedness speaks loudly to a young but maturing age group entering a world to which they must adapt in order to thrive.[34]

Another substantial way in which youth gets fetishized in the *Shadowhunters* universe is through the sexualization of the show's major characters. The Clary Fray from the book series *The Mortal Instruments* entered the narrative at the age of fifteen, shy, awkward, and a bit mousey. Clary of the novels shares little in common with the eighteen-year-old, buxom, Titian-haired beauty from the *Shadowhunters* television series. This newer version of Clary is also bolder, more expressive, and deeply opinionated. Yet shifting her age from fifteen to eighteen also renders Clary legally consumable as an object of desire.

Clary is not the only character who is positioned as an object of consumable desire within the series. Indeed, as critic Cheryl Eddy noted, *Shadowhunters* is a program "ostensibly about supernatural adventures, but actually about getting as many insanely good-looking young people into a single shot as possible."[35] Jace is blond, muscled, with golden, rune-covered skin (which viewers see quite a lot of), as well as a full lipped, pouty smirk and cocky attitude which make him the stereotypical bad boy; Alec, whose brooding, swarthy good looks and stern attitude toward duty intermix with his growing sexual confusion, rendering him both strong *and* vulnerable; while Isabelle, who dons a platinum wig going into battle because "demons dig blondes,"[36] constantly poses for the camera, casts seductive glances at both friends and enemies, and clads herself in clothing that maximizes the visibility of both her olive skin and her curvaceous figure beneath. Even Clary's best friend Simon, who is characterized as shy, awkward, and a bit geeky, receives an ample amount of shirtless airtime and in one scene is described by Isabelle as "nerd-hot."[37]

Likewise, many of the adults within the series exhibit a youthfulness and physicality that foils any reading of them as purely parental figures. Hodge, the former Circle member and mentor to the younger Shadowhunters, is tall and thin, with "gray-streaked hair and a long beaky nose" in the *City of Bones* novel,[38] yet on TV he is blond, muscular, fit, and a formidable martial arts master who fights shirtless through most of his airtime. Similarly, Luke

(Jocelyn's boyfriend and Clary's surrogate father) is described in the book series as dressed in "old jeans, a flannel shirt and bent pair of gold-rimmed spectacles that sat askew of his nose" with eyes "very blue behind the glasses."[39] Luke of the novel series as well as in the film adaptation *The Mortal Instruments* is clearly Caucasian. However, in the TV series Luke (played by Isaiah Mustafa, star of the oddly surreal and homoerotic series of Old Spice commercials) is African American, youthful, well built, and continually attired in tight-fitting jeans. Like Hodge and the young male leads, Luke is remarkably fit and spends a considerable amount of time with his shirt off as well.

Such images of youth, youthfulness, desirability, and unveiled sexuality infiltrate not only the *Shadowhunters* characters, plot lines, and overall message, but also the visuals and sales tactics of sponsored advertisements and other network programming, and reinforce the network's overall appearance and mission. With a target demographic of 14 to 34, an audience for whom the network coined the term "Becomers,"[40] Freeform's audience ranges in age from adolescents to adults. The network's performance in appealing to this audience seems to rest on validating and venerating its lower age tier of Becomers while at the same time appealing to the desires of its older viewers to visually consume images of youth.

Shadowhunters' ratings with this audience appear substantially lower than Freeform's network and cable TV counterparts. As of April 30, 2017, Freeform's top-rated program, *Pretty Little Liars*, averaged 1.19 million live viewers in the 18 to 49 age range, whereas *Shadowhunters* had an average of .70 million. For comparison, shows such as CBS's *The Big Bang Theory* (2007–present) routinely draw in over 14 million viewers from the same 18 to 49 demographic. While Freeform's appeal could definitely be considered *boutique* in that it appeals to a specific audience—the first twenty-year chunk of millennials—it is important to keep in mind the viewing habits of this particular demographic. According to a recent article from *International Business Times*, millennials "aren't watching television," at least not in the same way as their predecessors.[41] Data for this group from 2015 demonstrated a trend toward "nontraditional viewing" such as video on demand, network websites and apps, as well as streaming services like Hulu, accounting for 30 percent of millennial viewership.[42] Of these means for watching televisual content, Hulu topped the list of millennials' most go-to option. Keeping in mind that Freeform streams new episodes of *Shadowhunters* on Hulu as they air, as well as on the networks' own standalone app, it becomes difficult to truly establish the popularity of the show based on live TV ratings alone. Nielsen is now using a system they call "total audience measurement" which counts "all views, across all platforms, across a much longer period."[43] However, it takes considerably longer to establish ratings with this system. Therefore, it can be misleading to tabulate *Shadowhunters'* success based on live rating tallies alone. Given that Freeform chose to renew Shadowhunters based on the 2016 ratings, it seems clear the network is taking Hulu and other viewing options into account when determining overall success.

YA Fiction and Millennials

The YA literary genre made its debut in 1942 with the publication of Maureen Daly's *Seventeenth Summer,* although the actual genre term "young adult" didn't come into being until the 1960s, when the Young Adult Library Services Association coined it "to represent the 12–18 age range."[44] Authors from the early days of YA include S. E. Hinton, Judy Blume, Lois Duncan, and Robert Cormier, while the so-called "golden age" of YA peaked around the 1980s and then declined significantly during the 1990s. The genre's popularity reemerged around 2000, during which time the "second golden age" commenced.[45] At this point publishing houses began to appeal directly to teens, using the YA label as a marketing tool as much as a means for identifying the genre.

YA presently finds it target audience in the 12–18 age range, yet current research data reflects that adults in their 30s and 40s consume YA fantasy fiction like *Shadowhunters* as voraciously—or as some studies show, even more so—than their younger counterparts.[46] There are those in the literary world who believe it ridiculous for adults to consume YA literature, let alone to watch it on television. Ruth Graham's infamous screed "Against YA" argues that adults should feel embarrassed reading fiction written for teenagers. Graham cites a 2012 survey that found "55 percent of these books are bought by people older than 18," with the largest percentage of buyers being "between ages 30 and 44." That these books are geared toward teens aged 12–18 is what Graham sees as problematic.[47] Graham feels that by consuming YA fiction, adults are selling themselves short culturally and engaging in regressive, youth-yearning behavior. Many dissenting voices arose in opposition to Graham's stance, arguing that personal taste in literature should be private business.[48] At the same time, her opinion touches on the youth fetishism and consumption issue from Rob Latham's analysis, in that it demonstrates the increasing rate at which so-called adults fetishize the trappings of youth, potentially at the peril of losing contact with those elements of culture that once separated adults from children.

Yet such an opinion rests on the assumption that one cannot critically engage with youthful subject matter after a certain age without succumbing to a sort of dumbing down of the intellect. Additionally, Graham presumes that adults who engage with YA are not simultaneously reading and viewing other forms of literature, art, and entertainment. For instance, one can read

John Green's 2012 YA novel *The Fault in Our Stars* one day and Homer's *Odyssey* the next. The consumption of one form of literature does not negate access to all others. Further, how might Graham's opinions extend to other fiction genres generally considered to be of an escapist nature, such as adult detective novels, say, or science fiction and fantasy not marketed under the YA brand? How might these same critics feel about 34-year-olds viewing *Shadowhunters* as opposed to *NCIS* (CBS, 2003–present), *Game of Thrones* (HBO, 2011–present) or *The Americans* (FX, 2013–present)? Is it not possible to enjoy them all?

Much of the debate over YA fiction brings up the question of maturity and adulthood. The notion that maturity means growing up, getting a good job, and starting a family—the quintessential heteronormative narrative—elides the many other ways a person can grow into maturity. Given *Shadowhunters'* many flaws in terms of youth exploitation, this is one area where the show's creative team approaches the matter in a more nuanced fashion from other facets of the *Shadowhunters* narrative. Getting married and starting a family does not necessarily signify maturity from the young *Shadowhunters'* perspective. Indeed, Alec, the only character in season one to seek out a traditional heteronormative life path, changes his mind at the marriage altar, running off to be with Magnus instead. In this terrain, finding access to one's personal truth means something different for each character, and each does so in their own unique and inventive ways.

The white, patriarchal, heteronormative perspective that adulthood means coupling, procreating, and, in many ways, losing one's individuality in favor of a shared familial identity is anathema to those who choose less conventional routes. Critics such as Leslie Fiedler, whose views characterize "Twain's Huck Finn, Melville's Ishmael, and countless other canonical American literary characters [as] boys who refused to be civilized, who preferred a perpetual, homosocial boyhood to the responsibilities of adulthood—in particular the responsibilities of mature heterosexual relationships"[49] perhaps miss the point that mature heterosexual relationships aren't all there is to being an adult. At the same time, one might argue that our youth-obsessed, post–Fordist consumer culture has turned us all into adult-children, making us consuming vampires who feed on and off of youth, for whom the heteronormalized life of sacrifice, work, and family requires a growing up that is just not as much fun as playing all day long. However, this perspective seems a bit bleak and leaves little room for individuality or personal creativity in determining one's own life course.

This poses a series of questions: what exactly defines adulthood in the *Shadowhunters'* universe? Does an adult follow rules or rise above simplistic conformity? Does an adult make her own way in the world, no matter how unconventional the path, or does one follow the path set out for her by others?

Is rebellion always a childish act? If so, what of political uprisings in the name of social justice? Why is it that we assume growing up means falling in line with a pre-constructed, predetermined hierarchy? What of duty? When duty to the greater good comes in conflict with duty to the self, which is the more mature route?

This emphasis placed upon duty, purpose, and drive within the *Shadowhunters* narrative foils the oft-leveled accusation that millennials have no work ethic. While the eponymous characters supply plenty of youthful beauty and angst-driven melodrama to entertain the viewing audience, they also provide examples of young people who work as hard as they play, and each does so without embracing that traditional path of the so-called mature heterosexual relationship, but rather through personal inventiveness, ingenuity, and integrity. Of course, it is worth noting here that if we examine this conflation of youth and work through Rob Latham's vampire-cyborg lens, what we see is a system in which the youth of the series are becoming ever more cyborgized and absorbed into the economic machine that's been consuming youth for decades.

Chasing Shadows

Jean Baudrillard revisited Marx's use- and exchange-value model, finding that in contemporary consumer capitalist culture, rather than paying exchange-value prices for anything useful, we instead purchase desire.[50] Desire—or what W.F. Haug characterized as the promise of use-value absent any actual value of use—is, after all, what we end up buying when we aim to consume the reified construct of youth. Consuming desire sets up an endless and insatiable dynamic, a vortex of *want* that just *wants* more and more, the more it feeds on that *want*, the more *want* it accumulates, until one is eaten up by it all. Perhaps this is what it means that one must barter one's humanity to satiate desires for youth and longevity. At least that would be the conclusion if we look to the vampire or cyborg archetypes as images of extended youthfulness and longevity.

The *Shadowhunters* series provides a virtual space for engaging with youth and consuming its promises of vitality, inventiveness, and sensuality. The title of the show alone, the image of hunting for shadows, chasing phantasms through the street is so like the image of one endlessly pursuing the fleeting, illusory, and abstract idea of youth. And yet, chasing phantasms as we shop and consume feeds the voracious marketplace, itself a sort of mindless bloodsucking fiend, detached entirely of any human fleshliness. The abstract market feeds us abstract promises as we feed it abstractions of our labor in the form of money. Forever young is a wish we can embrace as we shop and as we consume our escapist fiction. But do we really want to live forever?

Notes

1. "The Mortal Cup," *Shadowhunters*, season 1, episode 1, Freeform, January 12, 2016.

2. There is a much larger discussion surrounding the publication and work of Cassandra Clare that, while very significant, belongs in a much larger conversation and is not connected to this essay. For more information, see Sabrina Rojas Weiss, "Everything You Need to Know About Cassandra Clare's Controversies," *Refinery 29*, January 13, 2016, http://www.refinery29.com/2016/01/100329/cassandra-clare-plagiarism-controversy, and Carolyn Cox, "*Dark Hunter* Author Sherrilyn Kenyon Suing Cassandra Clare Over *Shadowhunters*," *The Mary Sue*, February 12, 2016, https://www.themarysue.com/shadowhunters-lawsuit/.

3. The section is a summary of events that take place in the first season of *Shadowhunters*, supplemented with information from the *Shadowhunter's Codex* and the novel itself. See *Shadowhunters*, season 1, Freeform, January 12–April 5, 2016; Cassandra Clare and Joshua Lewis, *The Shadowhunter's Codex* (New York: Margaret K. McElderry Books, 2013), and Cassandra Clare, *City of Bones* (New York: Margaret K. McElderry Books, 2007).

4. A.O. Scott, "The Death of Adulthood in American Culture," *The New York Times*, September 11, 2014, https://www.nytimes.com/2014/09/14/magazine/the-death-of-adulthood-in-american-culture.html?_r=2.

5. Rob Latham, *Consuming Youth: Vampires, Cyborgs and the Culture of Consumption* (Chicago: University of Chicago Press, 2002), 30; W.F. Haug, *Critique of Commodity Aesthetics: Appearance, Sexuality and Advertising in Capitalist Society* (Minneapolis: University of Minnesota Press, 1886), 39–44, 87–94, quoted in Latham, *Consuming Youth*, 30.

6. W.F. Haug, *Critique of Commodity Aesthetics: Appearance, Sexuality and Advertising in Capitalist Society* (Minneapolis: University of Minnesota Press, 1886), 8.

7. Karl Marx, *Das Kapital: A Critique of Political Economy*, ed. Frederich Engels (Washington, D.C.: Regnery Gateway, 2000), 63–71.

8. Haug, *Commodity Aesthetics*, 16.

9. *Ibid.*, 41–42.

10. Latham, *Consuming Youth*, 3.

11. *Ibid.*

12. *Ibid.*, 27.

13. *Ibid.*, 5.

14. *Ibid.*, 5–6.

15. *Ibid.*, 4.

16. Clare and Lewis, *The Shadowhunter's Codex*, 147.

17. *Ibid.*, 23.

18. *Ibid.*, 36.

19. "Dead Man's Party," *Shadowhunters*, season 1, episode 3, Freeform, January 26, 2016.

20. Joyce Appleby, *The Relentless Revolution: A History of Capitalism* (New York: W.W. Norton, 2010), 258.

21. Jean Baudrillard, *For a Critique of the Political Economy of the Sign* (St. Louis: Telos Press, 1981), 83.

22. Baudrillard, *Political Economy*, 83–84.

23. Antonio Gramsci, *Selections from the Prison Notebooks*, ed. and trans. Quintin Hoare and Geoffrey Nowell Smith (New York: International, 1971), 285.

24. David Harvey, *The Condition of Postmodernity* (Cambridge: Blackwell, 1990), 125.

25. Keith Sward, *The Legend of Henry Ford* (New York: Rinehart & Company, 1948), 53.

26. Latham, *Consuming Youth*, 13.

27. *Ibid.*, 13–15.

28. *Ibid.*, 15.

29. David Harvey, *The Condition of Postmodernity* (Cambridge: Blackwell, 1990), 140–147.

30. Stuart Hall, "Brave New World," *Marxism Today* (October 1988): 24.

31. *Ibid.*

32. Latham, *Consuming Youth*, 16.

33. "Raising Hell," *Shadowhunters*, season 1, episode 4, Freeform, February 2, 2016.

34. Christopher Munsey, "Emerging Adults: The In-Between Age," *American Psychological Association* 37 (2006): 68; Jeffrey Jensen Arnett, "Emerging Adulthood: A Theory of Development from the Late Teens Through the Twenties," *American Psychologist* 55, no. 5 (2000): 469–749; Janel E. Benson and Glen H. Elder, "Young Adult Identities and Their Pathways: A Developmental and Life Course Model," *Developmental Psychology* 47 (2011), 1646–1657. Recent studies divide those in the 18–24 age group into what psychologist Jeffrey Arnett describes as emerging adults: no longer adolescents, but not yet established in their adult lives. Among the common traits shared by this age group are an ongoing exploration of personal identity and feelings of instability and of being in between childhood and adulthood, but not quite situated in either one. At the same time, this group is characterized by a sense of optimism that they will secure better professional and personal lives than their parents, while simultaneously more likely to engage in high risk behaviors such as, unprotected sex, substance use, and driving while intoxicated.

35. Cheryl Eddy, "The Next Season of Shadowhunters is Adding More Shadows, Which Will Fix Everything," *Gizmodo*, September 30, 2016, http://io9.gizmodo.com/the-next-season-of-shadowhunters-is-adding-more-shadows-1789522323.

36. "The Mortal Cup," *Shadowhunters*, season 1, episode 1, Freeform, January 12, 2016.

37. "The Decent Into Hell Isn't Easy," season 1, episode 2, Freeform, January 19, 2016.

38. Cassandra Clare, *City of Bones* (New York: Margaret K. McElderry Books, 2007), 66.

39. *Ibid.*, 22–23.

40. Elizabeth Wagmeister, "ABC Family to Rebrand Network 'Freeform' in January." *Variety*, October 6, 2015, http://variety.com/2015/tv/news/abc-family-freeform-rebranding-network-1201610697/.

41. "From 'Shadowhunters' to 'Shannara,' YA Fantasy Beguiles TV Networks, but Will Young Viewers Follow?" *International Business Times*, January 8, 2016, http://www.ibtimes.com/shadowhunters-shannara-ya-fantasy-beguiles-tv-networks-will-young-viewers-follow-2256636.

42. "Millennials and TV: New Data Shows Young People Still Watching TV, but Not Always on TV," *International Business Times*, September 20, 2015, http://www.ibtimes.com/millennials-tv-new-data-shows-young-people-still-watching-tv-not-always-tv-2192385.

43. *Ibid.*

44. Ashley Strickland, "A Brief History of Young Adult Literature," *CNN*, April 15, 2015, http://www.cnn.com/2013/10/15/living/young-adult-fiction-evolution/index.html

45. *Ibid.*

46. "Young Adult Books Attract Growing Numbers of Adult Fans," *Bowker*, September 13, 2012, http://www.bowker.com/news/2012/Young-Adult-Books-Attract-Growing-Numbers-of-Adult-Fans.html; "New Study: 55% of YA Books Bought by Adults," *Publishers Weekly*, September 13, 2012, http://www.publishersweekly.com/pw/by-topic/childrens/childrens-industry-news/article/53937-new-study-55-of-ya-books-bought-by-adults.html; Valery Peterson, "Young Adult and New Adult Book Markets," *The Balance*, March 17, 2017, https://www.thebalance.com/the-young-adult-book-market-2799954.

47. Ruth Graham, "Against YA," *Slate*, June 6, 2014, http://www.slate.com/articles/arts/books/2014/06/against_ya_adults_should_be_embarrassed_to_read_children_s_books.html.

48. Julie Beck, "The Adult Lessons of YA Fiction," *The Atlantic*, June 9, 2014, https://www.theatlantic.com/entertainment/archive/2014/06/the-adult-lessons-of-ya-fiction/372417/; Noah Berlatsky, "Young Adult Fiction Doesn't Need to Be a 'Gateway' to the Classics," *The Atlantic*, October 17, 2014, https://www.theatlantic.com/national/archive/2014/10/young-adult-fiction-doesnt-need-to-be-a-gateway-to-the-classics/381959/; Elisabeth Donnelly, "Slate's Condescending 'Against YA' Couldn't Be More Wrong—Young Adult Fiction Is for Everyone," *Flavorwire*, June 6, 2014, http://flavorwire.com/461021/slates-condescending-against-ya-couldnt-be-more-wrong-young-adult-fiction-is-for-everyone.

49. Christopher Beha, "Henry James and the Great Y.A. Debate," *The New Yorker*, September 18, 2014, http://www.newyorker.com/books/page-turner/henry-james-great-ya-debate.

50. Jean Baudrillard, *Selected Writings*, ed. Mark Poster (Stanford: Stanford University Press, 2001), 2.

Pretty Little Feminist Friendships

Pretty Little Liars' *Role* *in Deconstructing the Mean Girl Myth* *While Supporting Sisterhood*

ERICA LANGE

Representations of girls' and women's friendships on television depict how women in close relationships interact with one another in different stages of life, often with focus on one specific range within girlhood, adolescence, or adulthood. The show *Pretty Little Liars* (ABC Family/Freeform, 2010–2017) started as a teen-centered drama about a group of young female friends before the show went through a five-year time jump, carrying the storyline into the adulthood of the characters. The shift in time provides viewers a unique look at a friend group dynamics in two ranges of time: adolescence and adulthood. The portrayals spanning two intervals are significant because it broadens the scope of televised friendship models to include enduring bonds. Television friendships influenced my conceptualization of a best friend, and may similarly impact other viewers as well. Like many women, I grew up watching portrayals of best friends on television. Two of the most influential pairings for me were DJ (Candace Cameron-Bure) and Kimmy (Andrea Barber) on *Full House* (ABC, 1987–1995) and Blossom (Mayim Bialik) and Six (Jenna von Oy) on *Blossom* (NBC, 1990–1995). Watching best friend duos created an expectation that I was supposed to have a best friend too, which caused conflict for me when I realized I wanted to have more than one close friend. The relationships I saw on television made me feel that I was supposed to have a singular confidant, or one friend that should be better or more valuable to me than other friends.

The notion of a best friend pair is consistent throughout television history, with the most popular twosome showcased in the early 1950s with Lucy (Lucille Ball) and Ethel (Ethel Mertz) on *I Love Lucy* (CBS, 1951–1957). The solitary best friend notion maintained visibility and popularity into the late 1970s with *Laverne and Shirley* (ABC, 1976–1983), and is still widely accepted. Currently, we see portrayals of female best friendships on popular television shows such as Rory (Alexis Bledel) and Lane (Keiko Agena) or Sookie (Melissa McCarthy) and Lorelai (Lauren Graham) on *Gilmore Girls* (The WB, 2000–2007), Serena (Blake Lively) and Blair (Leighton Meester) on *Gossip Girl* (CW, 2007–2012), Leslie (Amy Poehler) and Ann (Rashida Jones) on *Parks and Recreation* (NBC, 2009–2015), and Abbi (Abbi Jacobson) and Ilana (Ilana Glazer) on *Broad City* (Comedy Central, 2014–present)—just to name a few.

These relationships shape ideas for women of what close female friendships should look like. However, there are few shows that extend the positive depictions of friends beyond that of a best friend duo. *Pretty Little Liars* is one of the outlying examples of a circle of empowering friends. In this paper, I examine how television shows influence and reinforce commonly held beliefs surrounding female friendships, with a focus on the singular best friend as well as the "mean girls" representation of group dynamics. I address essentialist views toward female roles and how the expectations therein further influence ideas of friendship, along with the complications husbands/partners may place on female bonds. I argue the show *Pretty Little Liars* disrupts the solitary best friend and mean girl narratives, while promoting feminist friendships.

Because of What We See on TV

As previously mentioned, my early wish to establish a best friend was shaped by the programs I watched in my youth. There were several girls around my age in my neighborhood, but I was allowed to spend more time with one girl who lived across the street. We both had older sisters close in age, and it was likely convenient for our mothers to balance play schedules together. Regardless of the convenience factors, she became my best friend, and I had someone with whom I could emulate the best friendships I saw on television. The experience of learning to desire a singular best friend is a common narrative among many women. Carter Rees and Greg Pogarsky state that the concept of friendship is recognized when children are very young, and that children learn perspectives and empathy from their friends.[1] Further, they identify that children categorize friendships to distinguish between regular friends and a best friend,[2] and that "few adolescents report not having a best friend."[3]

The idea of best friendship remains with us in how we understand friendships as adults. Gilbert A. Churchill and George P. Moschis claim a generally held belief among behavioral scientists is that childhood experiences influence adult behavior.[4] Television shows, especially to adolescents, create an idea of what situations in life may include, and the distinction of fiction and life experiences may seem indistinguishable. When we conceptualize television as an extension of life scenarios, we become socialized through the messages received from the shows we watch. Churchill and Moschis address this "consumer socialization" in adolescents and situate it through Orville Brim's framework of life-cycle [age] socialization and state: "The criteria relevant to the functioning in any given social system are prescribed by that society. They are based on normative theories of human behavior and, in a sense, are efforts on the part of some members of that society to regulate the behavior of other members so that certain 'desirable' consequences follow."[5] While their concept of the regulation of societal norms was discovered decades ago, the practices remain. Society pushes for traditional ideas of normalcy, even when we have changed to the point that not everyone shares the same ideals. Yet, we seem to be unable to break free of comparing our experiences to televised relationships, even if only in terms of relatability. Much in the way advertisements are effective in making people wish for products, television shows also create social desires for viewers.

The influence of popular shows expands to inform how women construct and negotiate identities, including what it means to be a friend. Media scholar David Gauntlett states that on average, people in the United States watch about 3 to 4 hours of television per day, providing audiences information that affects our life experiences. He claims the media shows us relationships and situations from multiple perspectives. These outlooks model scenarios that may be outside of audience members' worldviews, but contribute to their understanding of society. Gauntlett continues to say that the media depictions we consume become an audience's main point of reference for human interaction, and specifically attributes television as a primary source in understanding the general concept of friendship.[6] The impact of television on friendship identity formation extends to the number and quality of friendships women may desire, as well as our views on best friend relationships.

Though the connection of best friendship may seem empowering for female bonding, the representations are often misleading because the message many women receive is that women should seek to have *one* best friend rather than multiple female friendships. *Pretty Little Liars* deviates from the notion of singular best friendships. The show, in addition to being a drama, is a thriller with a villain who haunts, terrorizes, and manipulates a friend group into perilous situations by threatening to expose their secrets, which had been shared in private confidence. Since the villain, known only as A, seems to be privy to the girls' deeply held secrets, A threatens to disrupt the main circle of friendships for the four (eventually five) women involved. The friend group deviates from the popular friendship duo representations because the girls do not pair off into two sets of best or better friends, and each friend is given equal value within the group. The representation of multiple friends of equal value disrupts the singular best friend expectation by offering an example showing multiple "best" confidants who appear more like sisters, or a best friend family.

Mean Girl Myth

It is unclear if the privileging of best friend pairs in popular television is a coincidence or if it is intended to discourage women from engaging in groups. The popularity of pairs could be attached to keeping production costs down from fewer contracts to pay, but it could also be part of a systematic oppression of the female voice. Alison Bechdel is accredited with highlighting the primary representations of women in pairs in popular culture from a comic she published in 1986 titled "The Rule."[7] The principles described in "The Rule" have more commonly become known as the Bechdel Test, or the Bechdel-Wallace Test. Initially, the test examined the role of women in movies, though it is now being applied more broadly to television, social media, and video games. The Bechdel-Wallace Test is simple, consisting of three rules. The first is that the movie or television show must have at least two women in the cast. Second, these women need to talk to each other. And third, their conversation needs to be about something other than a man.[8] As of April 2017, according to "The Bechdel Test Movie List," out of the 7113 movies in the database, 4096 (57.6 percent) pass all three of the rules, 725 (10.2 percent) pass two rules, 1552 (21.8 percent) pass one rule and 740 (10.4 percent) pass none of the rules.[9] The movie list is limited to women who are named in films, and the films are selected for the database through the Internet Movie Database (IMDb) listing of the top 250 movies per year. Granted, the list references movies and not television shows, but the test can be used to assess visible media depictions of women based on their relationship with one other woman, without discussing a man.

Measures like the Bechdel-Wallace test outline dominant representations in popular media that discourage women from connecting with each other and reinforce negative stereotypes. When a large number of women in popular media are presented as only speaking to another girlfriend about men, viewers receive the message that women should strive to be attached to a man, and her best friend can help lament the difficulties involved in securing/maintaining a spouse. However, *Pretty Little Liars* offers audiences an alternative view. The show fosters the idea of a group of friends who talk about their goals, careers, and more than just the boys they like. The friend group consists of Alison (Sasha Pieterse), Aria (Lucy Hale), Spencer (Trioan Bellisario), Emily (Shay Mitchell), and Hanna (Ashley Benson), who are not showcased as being in competition with each other for a man's attention, even when Hanna and Spencer's dating circles overlap with Caleb (Tyler Blackburn). Hanna and Caleb dated and broke up.[10] Eventually Caleb and Spencer date,[11] but break up later in the series.[12] Shortly after, Hanna and Caleb get back together.[13] Throughout the breakups and makeups, Spencer and Hanna handle the relationships maturely. They do not compete for Caleb's affection, and maintain their friendship together.

Because the positive representation of best friends is more common than that of women in groups, how we come to understand circles of women through television can become skewed. The typical front-runners for groups of women on television are *The Facts of Life* (NBC, 1979–1988), *The Golden Girls* (NBC, 1985–1982), *Designing Women* (CBS, 1986–1993), *Sex and the City* (HBO, 1998–2004), *Desperate Housewives* (ABC, 2004–2012), and *Girls* (HBO, 2012–2017). These shows seem to feature empowering circles of female friends; however, the conversations rarely drift outside of the men they like, or the shows depict collectives of women as sites where deception and cruelty occurs.

Shows like *Desperate Housewives* often portray groups of women as being in competition with one another, casting judgment on everything ranging from relationship status, to cooking abilities, to parenting skills. These representations shape and reinforce the idea that groups of women are potentially mean girls. The mean girl trope can be traced back to the late 1970s, where groups of women were depicted unfavorably in the media. The representations typically share the commonality of having a leader who influences other women to gain popularity. We have seen characters like Chris Hargensen (Nancy Allen) in *Carrie* (1976), Brenda Walsh (Shannen Doherty) in *Beverly Hills, 90210* (FOX, 1990–2000), Chanel Oberlin (Emma Roberts) in *Scream Queens* (FOX, 2015–2016), among others, to perhaps the most notable, the iconic Regina George (Rachel McAdams) from *Mean Girls* (2004). Granted, we do have counter narratives with shows like *Girls* and *Friends* (NBC, 1994–2004), but as Margaret Tally states, "we've simply become unused to seeing popular dramatic entertainment that takes female friendship seriously as a focus and preoccupation."[14] If groups of women are largely portrayed as damaging, women are more likely to stick with the safety of their best friend and not attempt connections with other women.

On the surface, *Pretty Little Liars* seems to be another show about mean girls. The title alone casts the group as deceptive by immediately labeling them as liars, insinuating they are untrustworthy. When audiences begin to watch the series, they identify the initial mean girl as Alison, who at first embodies the trope as the leader of the group. Alison is pretty and popular, and she binds the friend circle together by knowing their secrets. The commonality of and reasons for mean girl behavior are analyzed by Dawn H. Currie, Diedre M. Kelly, and Shauna Pomerantz, who research schoolgirl meanness as a form of agency. Examining the role of mean girls in navigating the conflicting identity markers faced by women, the authors claim: "Girlhood as a culturally constructed 'way of being' is regulated by conventions that girls must be pretty but not 'self-absorbed' about their appearance; they must be attractive to boys but not seen to be too sexually 'forward'; they must be noticed and liked by the 'right people' but not a social climber; independent but not a 'loner'; and so on. Girls' agency therefore comes from a culturally mandated formation of girlhood."[15] Through this lens, meanness is used as a way to compete with other women for social status. By exposing girls as being in categories they "should not" be in, girls are able to position themselves as being more socially acceptable than other girls, often rewarded with popularity for acting as they are "supposed to." The increasing popularity provides a feeling of being heard by peers. The trouble is, this teaches women that in order to have their own social value, it must come at the expense of someone else's.

The concept of social gains is further expressed in *Pretty Little Liars* through the initial depiction of Alison. She uses the information the girls share with her to form a clique, encouraging the girls to continue sharing their secrets, saying that is what close friends do.[16] Alison uses what is shared confidentially to threaten the group and keep an enclosed circle of friends. Being part of Alison's clique enables the group to be the most popular girls in school, but their secrets are a weapon that can be used for social exclusion.

When Alison goes missing in the pilot episode of *Pretty Little Liars*, the girls talk about feeling a sense of relief. They identified her meanness and welcomed the opportunity to explore their own individual identities, free of the fear of social exploitation. However, the relief is short-lived, as another mean girl steps in Alison's place. This time, the mean girl appears invisible and seemingly everywhere at once, sending threats via text message while going only by the letter A. The secrets that bound the girls together no longer signify a closeness between friends. As A threatens to expose things only the girls and Alison could know, they are manipulated into maintaining contact with each other and carrying out the demands of A. The show then shifts from a whodunit to a who*isit*. The group becomes tighter as they realize they must depend on one another to manage the manipulations of A, and also because the existence of A is a secret they cannot tell.

Seeing examples of women manipulating other women on television shows perpetuates negative stereotypes of how women interact with one another. Gloria Cowan and Jodie B. Ullman cite feelings of inadequacy as a predecessor of stereotyping, and position both as sources of women's hostility toward women. They state that "women who have a sense of inadequacy are more likely to project these negative feelings onto other women, making other women the targets of their personal malaise."[17] Additionally, Cowan and Ullman argue that "the devaluation and trivialization of women is likely to be internalized to some extent by most women…. [This] social devaluation of women may create the propensity for women to devalue women as a group."[18] When we feel the need to assess a level of risk in terms of stereotyping or hostility when socializing with other women, we become guarded in what we are willing to share, leading us to miss out on opportunities to connect and formulate bonds. The examples of manipulation encourage women to silence themselves and not discuss things that can cast them in a bad light. While manipulation was what initially connected the *Pretty Little Liars* group, as the series progresses, the friendships do not become disenfranchised. Even Alison shifts from being the number one mean girl to becoming a valued member and confidant of the circle.

Viewers see a much different version of Alison upon her return to Rosewood. During her marriage to Elliott Rollins (Huw Collins), she openly discusses her struggle with mental health and asks Emily to take her to check into The Radley Sanitarium.[19] Oftentimes, admitting the need for psychological help could lead to public shaming, and mental health has often been seen as something that should not be discussed in order to avoid stigmatization. In addition to Alison's openness about wanting help, *Pretty Little Liars* functions to destigmatize treatment, and positions it instead as something acceptable that may be encountered in one's life. We see this not only with Alison, and Spencer, but through several other female characters including Mona (Janel Parrish), Charlotte (Vanessa Ray), Marion Cavanaugh (Karla Droege), and Mary Drake (Andrea Parker), who were also admitted to Radley at various points in the series.

When the mean girl trope is dismantled by Alison's disappearance, the opportunity for a band of girlfriends surfaces and remains consistent throughout the series. The group is shown discussing matters of health, goals, and even stage an intervention when Spencer abused pills to achieve her academic goals.[20] The friends continue to be a support system for one another, and upon Alison's return,[21] she enhances the sisterhood by revealing that while she was presumed dead, she had been looking out for the group without their knowledge. She does not reassume her former mean girl role. When Alison confides in Emily during her rehabilitation process, she demonstrates a true connection that is not reliant on her former practice of coercion.

Mendacity of Mon-A

Mona complicates the mean girl myth on the show. She starts out as a nerdy character who clearly wants to fit in with the popular girls. After Alison's disappearance, she befriends Hanna and even addresses the rest of the *Pretty Little Liars* squad as "her friend's other friends." Soon after, Spencer discovers Mona is working as an early version of A terrorizing the girls, betraying Hanna's trust. However, it is revealed that Mona is not *the* A, but was part of a ring of As coerced into doing the dirty work of the real A, whose identity no one involved actually knew. Nonetheless, Mona's involvement as an A ensures that she will never fully gain the trust of the group. Despite the girls' dislike and distrust of Mona, they find themselves in situations where they have no option but to reluctantly put faith in her from time to time. Mona's desperation to be included by the group is evident throughout the series, but the only way she only ever accomplishes this is through forcing herself to become an ally with the group against the real A after deceiving them. Cameron Anderson and Gavin J. Kilduff assert that "individuals pursue status by enhancing the apparent value they provide to the group."[22] Additionally, they claim individuals gain or rise in status within a group mostly through "acting in ways that signal task competence, generosity, and commitment to the group."[23] Likewise, though Mona is aware she will never wholly gain the acceptance of the group or become an insider, she maintains her connection to the women in some way because it is the closest thing to friendship she knows. However, even with Mona's numerous attempts to help the group, she never fully recovers socially from the damage she caused as A[24]; in the series finale, she ends up without a friend group, yet appears satisfied with trapping Mary Drake and Spencer's evil twin, Alex Drake (Troian Bellisario) in her own dollhouse, allowing them to function as both prisoners and entertainment.[25]

This portrayal of Mona is representative of the complicated struggle women experience when faced with "frenemies," or women who appear to be friendly on the surface, but secretly feel they are in competition with other women, or are otherwise disingenuous. The idea of a frenemy is not certainly exclusive to Mona's character, nor to *Pretty Little Liars*. While the best-known example of frenemies may be the young women in *Mean Girls*, we see another frenemy relationship with Jane (Gina Rodriguez) and Petra (Yael Grobglas) on *Jane the Virgin* (The CW, 2014–present). Both women appear friendly and respectful to each other, despite a distrust in how they perceive one another's relationship motives for Rafael (Justin Baldoni), who has fathered children with both of them. The frenemy trope enhances the notion that women can be cruel to one another, while still seeming to be friends. Though the frenemy concept is similar in both shows, Mona tries to detach herself from being labeled as untrustworthy by offering help to the group. Though the girls are bound together by secrets, their trust in one another becomes sacred. It is necessary to observe how trust operates within the *Pretty Little Liars* circle, because it shows viewers that women in groups are not always cruel and *can* be trusted, while simultaneously presenting a model of how to set healthy boundaries that do not include the notion that our trust should be extended to all women we encounter.

Going Along to Get Along

Women on television have a history of being portrayed as caring and nurturing, with these qualities usually ascribed to mature or motherly women like June Cleaver (Janine Turner) on *Leave It to Beaver* (CBS, 1957–1963). The nurturer role pushes women into what I refer to as the "mommy box," or the positioning of women's ultimate role to act motherly and caring to a fault. Kathleen M. Ryan and Deborah A. Macey claim television subtly shapes women's expectation of motherhood, creating a societal narrative indicative of what women should feel and think about the roles of their gender, pushing the notion that motherhood should be a woman's ultimate goal.[26]

Mark Barner's assessment of essentialized representations of women in television shows revealed that women are underrepresented in children's programs, and that men and women in these shows are highly stereotyped. He states, "The typical male character makes and carries out plans, is active, dominant, aggressive, and seeks attention…. The typical female character is deferent, dependent, and nurturing. Also … she is ignored much of the time."[27] These characteristics are often essentialist views ascribed to female and male identities, and internalized by viewers from a very early age. Barner positions television as highly influential in socialization and states that television is "arguably the most popular, constant, and consistent source of information on, among other things, socialization, including that which is expected, accepted, and taken for granted."[28] As such, the essentialist views expressed and reinforced through the roles of women on television have become normative stereotypes prevalent in American culture. Similarly, K. Anthony Appiah argues that stereotypes shape how we form identity conceptualizations, including those attached to gender. He states stereotypes are "grounded in a social consensus about how [women] ought to behave in order to conform appropriately to the norms associated with membership in their group."[29] The idea that being caring and nurturing are female identity norms conflicts with the mean girl representations of women in groups.

Television tends to separate women's roles primarily into two categories: mean girls and nurturers. In addition to the normalized stereotype that *acceptable* women do not speak up, nurturers face further constraints of appropriateness in terms of maturity and immaturity. Jane Rosenzweig noted the trend of television infantilizing single adult women on shows like *Ally McBeal* (FOX, 1997–2002), *Wasteland* (ABC, 1999), and *Providence* (NBC, 1999–2002). She uses teenage dramas like *Dawson's Creek* (The WB, 1998–2003), which formed a cornerstone for the popularity of teen-centered television, to argue that the shows' high ratings may have influenced the portrayal of single womanhood as an extension of adolescence.[30] This, in turn, created a delineation of maturity and immaturity in how we view single and married women culturally. The idealized need to nurture presented through the mommy box has become the socialized standard of maturity. Movies like *Preggoland* (2015) and *La La Land* (2016) show women maturing into adulthood through parenting, reinforcing the notion that acting motherly or motherhood is deemed as what makes women mature, and causes women who do not aspire to motherhood to be viewed as immature.

Upon Alison's return to Rosewood as an adult, she is married, though her husband is out of the picture. After she is revealed to be pregnant, Allison is rarely seen outside of work and home. However, she does not conform completely to the mommy box trope as she struggles with her decision to have the baby. Allison is unaware that her pregnancy is not the result of her husband, at least in the traditional sense. When she was admitted to Radley, Alison's husband, Elliott, was discovered to be part of the manipulative A.D. (the Uber A and villain of later seasons) team, and he drugged Allison to perform an artificial insemination using stolen eggs that Emily had donated during a time of financial difficulty. When A.D. informs Alison of the devious circumstances of her pregnancy, she consults with Emily and ultimately decides to not go through with having an abortion.[31] Because the mommy box has become synonymous with maturity, Allison's transition into motherhood is the solidifying signal that she is officially a mature adult and no longer a mean girl. When being caring and nurturing becomes accepted as the binary opposite of mean girl behavior, both ends of this stereotype spectrum are detrimental to women. Maturity becomes viewed as the ability to tolerate negativity with grace, and while this does not sound bad, it normalizes the idea that women should ignore being treated poorly because being polite is believed equal to being mature. This is illustrated throughout *Desperate Housewives*, specifically when Lynette Scavo (Felicity Huffman) attends her first PTA meeting at Barcliff Academy and speaks up about the director's intent to rewrite the school play. Her comments are met with passive aggression that attacks her involvement in the production, including other parents telling her to leave the big decisions to the women who show more effort. Lynette smiles politely and refrains from further discussion of her opinion.[32] Maturity in Alison and Lynette's situations prohibits women from pursuing their own desires due to the fear of being cast as immature, or the potential of creating conflict.

The representations of immaturity and mean girls in opposition to maturity and nurturers form a false dichotomy wherein one representation is desirable or appropriate, and the other is inappropriate, culturally speaking. Though the media rewards mean girls by giving them preference and primacy, the negative feelings that viewers may take away from the characters become a basis for stereotyping and hostility. To that end, if women are not performing the "appropriate" ideal, they may be vilified for not being caring, nurturing, or nice "enough" and therefore be delegated to the undesirable parallel: a mean girl. The essentialized representations of women create an expectation that women should assume the role of peacemakers and nurturers, often at the expense of not speaking up and honoring feelings of hurt or distrust. This expected politeness devalues women in a way that subtly indicates it is acceptable to behave disrespectfully to women, and women should bite their tongues and be virtuous or risk being viewed as catty or immature.

Pretty Little Liars additionally subverts the notion of going along to get along as a marker of maturity through Mona's character by showing the girls standing up for themselves, even in situations where they are forced to rely on Mona to help them. They acknowledge their dislike for her and articulate why, emphasizing that their dislike is not a carryover from the initial mean girl scenario when they viewed her as "loser Mona," prior to Alison's disappearance at the beginning of the show. However, they have learned that Mona can be negative, or even toxic, and they do not accept her treatment politely. Instead of trying to act as peacemakers, they manage their encounters with her in ways that show viewers there is a difference between self-protection and being mean. Perhaps more importantly, the interactions demonstrate it is okay to step out of the mommy box and stand up for yourself.

Housewives and Social Lives

The mommy box construct is additionally problematic because it enhances the idea that women should desire a normative housewife role, often in opposition to pursuing a career. Women buy into what Colette Dowling coined "The Cinderella Complex," or the need for women to be dependent on men, due in part to a veiled fear of independence.[33] Concepts like the Cinderella complex have conditioned women to believe that they need a man in order to be happy, or to prove some ideal of worthiness or value. Additionally, this construct can imply that women should devote the majority of their time to a husband, or potential husband, and family life. Supporting this narrative, Sarah M. Flood and Katie R. Genadek claim it is common for couples to believe shared time is important for the well-being of a marriage.[34] Considering this, husbands may find their spouses having a singular best friend more acceptable than a group, as one close friend may be less likely to take too much of the matriarch's time away from the homestead.

Importantly, the Cinderella complex is not reinforced on *Pretty Little Liars*. Adult Hanna believed that A could cause harm to her fiancé Jordan (David Coussins), so she broke off her engagement with him.[35] In this way, Hanna became the protector, emphasizing that she does not need a Prince Charming. At the same time, Aria was contemplating whether or not to stay with Ezra when his former girlfriend Nicole (Rebecca Breeds) is found alive after being missing and presumed dead for many years.[36] Though Ezra stated several times that he wanted to remain engaged to Aria, she urges him to see his ex and does not demand he honor their engagement. Ultimately, Aria decides to stay engaged to Ezra when she is certain that he has moved on from Nicole, and not because she feels she has an obligation to marry him. These depictions encourage the notion that women can be selective in choosing marriage. The friend group also demonstrates that adult women are capable of building careers and maintaining friendships. This counter narrative positions women as sources of support that do not encourage dependence on a partner. Christine M. Proulx, Heather M. Helms, and C. Chris Payne analyzed studies which revealed "wives are more likely than husbands to engage their close friends in conversations about marital concerns and to seek out friends more than spouses for emotional support."[37] When there is greater opportunity to talk about things like independence or levels of happiness, multiple empowering friendships—like the representations on *Pretty Little Liars*—could perceivably threaten marriages by encouraging women to establish and achieve personal success or goals outside of the home.

Similarly, Proulx, Helms, and Payne argue that "female friends provide a support base for wives to confront their husbands on important issues such as wives' entitlement to time with friends."[38] A sisterhood of friends can offer help for women to seek independence if they are unhappy in a marriage, or provide a space for women to question or reevaluate their dependence on men. Though women may encounter challenges like time constraints with having children and partners preventing sisterhood friendships, presumably, the singular best friend is reinforced as the ideal friendship status for women, because certain oppressive partners may feel threatened by the prospect of multiple women spending time together.

Pretty Little Liars provides a nuanced perspective of understanding and questioning dependence on men when the friend group parses Alison's marriage to Elliot. After his true identity as Archer Dunhill is revealed, along with his shady past and manipulations of Alison,[39] the show offers a view of negative consequences of the Cinderella complex, showing that men are not always Prince Charming. Additionally, viewers see that the women do genuinely care for Alison, and provide a support system for her through a difficult time.

Pretty Little Empowerment

Pretty Little Liars displays another disruption of normative standards by portraying the group as strong and empowered. The girls are faced with situations of sheer terror, often in the dark. For example, when Aria finds herself in a crate with Officer Reynolds's (Yani Gellman) corpse on the Halloween episode in season three,[40] she is portrayed as brave and strong, despite being scared. Similarly, after A captures the group and keeps them in what became known as the dollhouse, the women are determined to find a way out.[41] When escaping, Caleb and Ezra (Ian Harding) arrive in time to help open the door, but the group has already managed most of the escape themselves. The women never hesitate to tackle a challenge or face a fear head-on. They are not portrayed as too fragile to fight, nor are they seen as victims. The viewer does not pity them in intense situations; rather, they cheer them on. Seeing the group as strong interferes with the "damsel in distress" notion we are fed in fairytales and provides an alternative to commonly held views that women need men to protect them. Yes, the group does get help from some of their partners, Caleb, Toby (Keegan Allen), and Ezra, but, the role of the men is not to swoop in and save the day, nor to save the girls, because they are shown to be perfectly capable of taking care of themselves.

When television shows fail the Bechdel-Wallace test or perpetuate the Cinderella complex, essentialized notions of gender inequality are reinforced and normalized, depicting women as fragile and dependent on strong men to "save" them from the perils of life. An indication of this trend can be identified in a study by Mark Brandt on sexism and gender inequality. Brandt looks specifically at how sexist ideologies contribute to gender inequality and found that "greater sexism predicts decreases in gender equality over time and provides the first direct evidence of sexism as a hierarchy-enhancing ideology."[42] His study also purports "that societal-level sexism predicts increases in systemic gender inequality."[43] Representations that encourage dependence on men or discourage spending time away from a partner are damaging to how women negotiate their perceptions of self-worth and desire, because as Julia T. Wood points out, "the point is not that having babies or committing to relationships is wrong; rather, it is that media virtually require this of women in order to present them positively[.] Media define a very narrow range for womanhood."[44]

Pretty Little Liars breaks away from the dominant gendered representations of women by showing the female leads wanting to be involved in romantic relationships, while illustrating that their partners do not define them. This is present through Alison's marriage to Elliott, which was based on him deceiving her. Alison goes on to make a life for herself outside of the experience of her marriage; she is shown focusing on her teaching career and being successful independent of Elliott, providing viewers with the notion that women do not have to tolerate bad relationships, and can move on from them.

Pretty Little Liars further broadens the scope of what is deemed acceptable surrounding the gendered constructs of being a woman as the group transitions from high school to young adulthood. The women are shown to be hard-working and career driven in the second part of season six as Hanna is pursuing a high-pressure career in fashion, Spencer has a job as a lobbyist on Capitol Hill, Aria is working in publishing in Boston, and Alison is a teacher. During this time, Emily is a bartender but goes on to become the swim coach at Rosewood High where Alison is teaching.[45] These representations allow viewers to see the empowerment of women making career and relationship decisions and provide an alternative take on the notion that women should desire the same housewife lifestyle.

Replacing Singularity with Sisterhood

My childhood friendship models and experiences encouraged me to have a singular best friend, but I learned to challenge this narrative as I got older, especially with relocations. As an adult, I have several "best" friends spread throughout different states. While some of my closest girlfriends have met each other, the only common denominator is me, and we do not form a social circle together. I realize that I should not have an expectation for all of my friends to be connected to one another, and I have met many women who feel apprehensive to engage with groups of women. In a 2010 study that supports this idea, Kelly Valen found that "roughly 60 percent of respondents [out of 3,020 women surveyed ages 15 to 86] say they still find themselves feeling uncomfortable, anxious, wary, awkward, cautious, intimidated, or even distrustful of other females as a result of past experiences."[46]

To negotiate feelings of anxiety and reshape the dominant discourse of female friendships, I offer a two-part suggestion: First, we should position our close friends as though they are our sisters. It has been my experience that women have less trepidation about the idea of engaging with a friend's family, or in my case, a friend family, as opposed to seeing a group of women potentially comprised of grown up mean girls. Psychologist Anne Campbell states, "there is a high degree of exclusivity in these [best] friendships and friends are acutely concerned with maintaining them. To spend too much time with a third girl is to threaten the bond between friends."[47] However, if the third girl is a sibling, women might be less inclined to perceive a friend's sister as weakening their friendship, or fostering feelings of exclusion. Perhaps if groups of friends are reframed and positioned as familial, the attachment to the term *best* will become deemphasized, but the friendship would not become diminished.

Second, I suggest moving from "sister friends" to (re)empowering sisterhood. Kelli Zaytoun and Judith Ezekiel state that "the Women's Liberation movement was predicated upon the idea of solidarity among women, once commonly referred to as 'sisterhood.'"[48] Additionally, Karen Hollinger positions ideas of sisterhood as being essential to the women's movement and elaborates, "The vision of a woman-identified community united by common interests and a determination to combat sexist oppression has fueled feminist rhetoric and ideological commitment."[49] She continues, "It should not be surprising, therefore, that this notion of sisterhood should have filtered into popular cultural representations aimed at a female audience."[50] As Zaytoun and Ezekiel point out, "Today the very term sisterhood seems somehow quaint, viewed by some as naïve or anachronistic, by others as ethnocentric and dogmatic."[51] In a time where women are forging bonds through venues like the 2017 Women's March on Washington, television and audiences need the type of sisterhood representations in *Pretty Little Liars* to gain greater visibility and reshape the dominant cultural view of women in groups as sites of cruelty.

The representations women see of close female friendships must resist the singular best friend narrative and enable women to conceptualize multiple positive close friendships. This is not to say that having a singular best friend should become undesirable, but rather to call instead for more representations of women building community and substantive connections, and fewer women in competition with one another. *Pretty Little Liars* is one of the few mainstream popular television shows that provides viewers a positive representation of a unified sisterhood aimed at supporting and empowering a circle of friends. The portrayals on the show are unique because they do not reinforce the idea that women should aspire to obtain a traditional housewife role, and they grapple with notions of sexism, stereotypes, and beliefs of what women should desire from life, as well as their friendships. Shows like *Gilmore Girls* offer similar sentiments in terms of female ambitions, but many of these shows have been cancelled and are only available in syndication, and may seem dated to a contemporary audience. With *Pretty Little Liars'* recent series finale, there is currently a need for more positive depictions of women on television that promote feminist friendships in the form of a sisterhood, portraying a positive community of women as something available, acceptable, and empowering for female audience members to engage in.

NOTES

1. Carter Rees and Greg Pogarsky, "One Bad Apple May Not Spoil the Whole Bunch: Best Friends and Adolescent Delinquency," *Journal of Quantitative Criminology* 27, no. 2 (June 2011): 197–223.

2. *Ibid.*, 198.

3. *Ibid.*

4. Gilbert A. Churchill and George P. Moschis, "Television and Interpersonal Influences on Adolescent Consumer Learning," *Journal of Consumer Research* 6, no. 1 (1979): 23.

5. Orville G. Brim, *Socialization through the Life Cycle* (New York: John Wiley & Sons, 1966), quoted in Gilbert A. Churchill, and George P. Moschis, "Television and Interpersonal Influences on Adolescent Consumer Learning," *Journal of Consumer Research* 6, no. 1 (1979): 23.

6. David Gauntlett, *Media, Gender and Identity: An Introduction* (New York: Routledge, 2008), 2.

7. Alison Bechdel, "Dykes to Watch Out For: THE RULE," *Off Our Backs* 16, no. 6 (1986): 27.

8. "What Is the Bechdel Test and Why Is It Important?" *Life Examiner* (USA), August 31, 2015.

9. "Bechdel Test Movie List," *Stats and Graphs—Bechdel Test Movie List*, http://bechdel test.com/statistics/.

10. "Who's in the Box?" *Pretty Little Liars*, season 4, episode 14, ABC Family, January 7, 2014.

11. "The Gloves Are On," *Pretty Little Liars*, season 6, episode 13, Freeform, January 26, 2016.

12. "Wanted Dead or Alive," *Pretty Little Liars*, season 7, episode 6, Freeform, August 2, 2016.

13. "The Darkest Knight," *Pretty Little Liars*, season 7, episode 10, Freeform, August 30, 2016.

14. Margaret Tally and Betty Kaklamanidou, *HBO's Girls: Questions of Gender, Politics, and Millennial Angst* (Newcastle upon Tyne: Cambridge Scholars, 2014), 28.

15. Dawn H. Currie, Deirdre M. Kelly, and Shauna Pomerantz, "'The Power to Squash People': Understanding Girls' Relational Aggression," *British Journal of Sociology of Education* 28, no. 1 (2007): 24.

16. "Pilot," *Pretty Little Liars*, season 1, episode 1, ABC Family, June 8, 2010.

17. Gloria Cowan and Jodie B. Ullman, "Ingroup Rejection Among Women: The Role of Personal Inadequacy," *Psychology of Women Quarterly* 30, no. 4 (2006): 406.

18. *Ibid.*, 399.

19. "Hush, Hush, Sweet Little Liars," *Pretty Little Liars*, season 6, episode 20, Freeform, March 15, 2016.

20. "Free Falling," *Pretty Little Liars*, season 4, episode 20, ABC Family, February 17, 2014.

21. "Did You Miss Me?" *Pretty Little Liars*, season 4, episode 13, ABC Family, October 22, 2013.

22. Cameron Anderson and Gavin J. Kilduff, "The Pursuit of Status in Social Groups," *Current Directions in Psychological Science* 18, no. 5 (2009): 295.

23. *Ibid.*

24. "Father Knows Best," *Pretty Little Liars*, season 2, episode 22, ABC Family, February 27, 2012.

25. "Farewell, My Lovely," *Pretty Little Liars*, season 7, episode 19, Freeform, June 20, 2017.

26. Kathleen M. Ryan and Deborah A. Macey, *Television and the Self: Knowledge, Identity, and Media Representation* (Lanham, MD: Lexington Books, 2015), 48.

27. Mark R. Barner, "Sex-Role Stereotyping in FCC-Mandated Children's Educational Television," *Journal of Broadcasting & Electronic Media* 43, no. 4 (1999): 561.

28. *Ibid.*, 551.

29. Anthony K. Appiah, "Stereotypes and the Shaping of Identity," *California Law Review* 88, no. 1 (2000): 48.

30. Jane Rosenzweig, "Ally McBeal's Younger Sisters," *The American Prospect* 11, no. 1 (November 23, 1999): 62–64.

31. "Power Play," *Pretty Little Liars*, season 7, episode 14, Freeform, May 9, 2017.

32. "Running to Stand Still," *Desperate Housewives*, season 1, episode 6, ABC, November 7, 2004.

33. Collette Dowling, *The Cinderella Complex: Women's Hidden Fear of Independence* (New York: Summit, 1981): 31.

34. Sarah M. Flood and Katie R. Genadek, "Time for Each Other: Work and Family Constraints Among Couples," *Journal of Marriage and Family* 78, no. 1 (February 1, 2016): 142.

35. "Bedlam," *Pretty Little Liars*, season 7, episode 2, Freeform, June 28, 2016.

36. "The Darkest Knight," *Pretty Little Liars*, season 7, episode 10, Freeform, August 30, 2016.

37. Christine M. Proulx, Heather M. Helms, and C. Chris Payne, "Wives' Domain-Specific 'Marriage Work' with Friends and Spouses: Links to Marital Quality," *Family Relations* 53, no. 4 (2004): 393.

38. *Ibid.*, 394.

39. "The Talented Mr. Rollins," *Pretty Little Liars*, season 7, episode 3, Freeform, July 5, 2016.

40. "This Is a Dark Ride," *Pretty Little Liars*, season 3, episode 13, ABC Family, October 23, 2012.

41. "Game On, Charles," *Pretty Little Liars*, season 6, episode 1, Freeform, June 2, 2015.

42. Mark J. Brandt, "Sexism and Gender Inequality Across 57 Societies," *Psychological Science* 22, no. 11 (2011): 1413.

43. *Ibid.*, 1416.

44. Julia T. Wood, *Gendered Lives: Communication, Gender, and Culture* (Stamford: Cengage Learning, 2001), 35.

45. "Playtime," *Pretty Little Liars*, season 7, episode 11, Freeform, April 18, 2017.

46. Kelly Valen, *The Twisted Sisterhood: Unraveling the Dark Legacy of Female Friendships* (New York: Ballantine Books, 2010): 86.

47. Anne Campbell, *A Mind of Her Own: The Evolutionary Psychology of Women* (Oxford: Oxford University Press, 2002): 155.

48. Kelli Zaytoun and Judith Ezekiel, "Sisterhood in Movement: Feminist Solidarity in France and the United States," *Frontiers: A Journal of Women Studies* 37, no. 1 (2016): 195.

49. Karen Hollinger, "'Respeaking' Sisterhood," *Studies in Popular Culture* 20, no. 1 (1997): 53.

50. *Ibid.*

51. Kelli Zaytoun and Judith Ezekiel, "Sisterhood in Movement: Feminist Solidarity in France and the United States," *Frontiers: A Journal of Women Studies* 37, no. 1 (2016): 195.

Negotiating Creative Feminine Labor on Family Television
Are Jane by Design *and* Bunheads *Riding a New Feminist Wave?*

JESSICA FORD

Bunheads (2012–2013) and *Jane by Design* (2012) both debuted in 2012 as part of ABC Family's (now Freeform) slate of original programing. Both series focus on non-normative families and explore creative labor. *Bunheads* follows dancer Michelle Simms (Sutton Foster) as she uproots her life as a dancer in Las Vegas, to eventually become a dance teacher in the small fictional seaside town of Paradise, California. *Jane by Design* centers on Jane Quimby (Erica Dasher), a high school student who goes for an internship and, through a series of misunderstandings, lands a job as an executive assistant to the creative director of fictional New York fashion designer Donovan Decker, an elusive God-like figure that is constantly discussed, but never appears in the series. In the pilot episodes of both series, the female protagonists' lives become exceedingly more complicated as they are ushered into their new worlds. *Bunheads* and *Jane by Design* each center on feminine creative arts (dancing and fashion) as their primary expression of self, femininity, and labor. Both series only ran for one season and have not been widely examined in academic or journalistic discussions of recent scripted U.S. television.

Bunheads and *Jane by Design* work within the larger programing tendencies of ABC Family, which leaned towards "family friendly."[1] Historically, the family drama is a rather traditional format with conservative family values and politics.[2] In recent years, however, ABC Family and Freeform have found

84

success with series that depict non-normative families and have progressive politics, such as *Switched at Birth* (2011–2017) and *The Fosters* (2013–2018). Interestingly, ABC Family's scripted programing often has a feminine bent, at least since the mid-2000s, with popular series like *The Secret Life of the American Teenager* (2008–2013), *Pretty Little Liars* (2010–2017), *Young and Hungry* (2014–present), and *The Fosters,* and less well-known series, including *Huge* (2010), *Bunheads*, and *Jane by Design*. These series are in keeping with the feminine flavor of their syndicated programing, which features reruns of *Gilmore Girls* (The WB/The CW, 2000–2007), *Reba* (The WB/The CW, 2001–2007), and *The Mindy Project* (Fox/Hulu, 2012–present). In her discussion of the ABC Family brand, Caryn Murphy argues that ABC Family heroines are not "passively feminine," but rather they are "independent, assertive," and they are "heroines who the audience is meant to root for."[3] While the feminine tendency of many of ABC Family's programs is well-acknowledged, less discussed is the network's feminist sensibility.

Despite certain similarities, *Bunheads* and *Jane by Design* are rather different series that operate in distinct ways, engage with various feminist and postfeminist ideologies, and occupy particular discursive spaces. For instance, as naïve and innocent as Jane is, Michelle is equally as knowing and worldly. *Bunheads* explores the intersecting lives of Michelle, a group of teenage ballet dancers, and their teacher, Miss Fanny Flowers (Kelly Bishop). *Jane by Design* follows high-school student Jane and her best friend Billy Nutter (Nick Roux), who both have absentee parents and as such they have formed their own de facto family with Jane's older brother Ben (David Clayton Rodgers). While *Bunheads* features teenage characters in Sasha (Julia Goldani Telles), Boo (Kaitlyn Jenkins), Ginny (Bailey De Young), and Melanie (Emma Dumont), it is not a teen soap opera in the tradition of *Dawson's Creek* (The WB, 1998–2003), *The O.C.* (Fox, 2003–2007), *Gossip Girl* (The CW, 2007–2012), or more recently *Riverdale* (The CW, 2016–present). Unlike the aforementioned teen soaps, *Bunheads*' narrative does not center on high school power dynamics, communal social events, illicit romantic affairs, or love triangles. *Jane by Design,* however, draws on the generic staples of teen soap more heavily, even going as far as to have many adult characters replay their high school dynamics as staff at the high school. As with the teen characters of *The OC* and *Gossip Girl*, the teenagers of *Jane by Design* are somewhat aspirational figures who are largely neglected by their parents. It is this parental neglect that enables the characters to operate as independent young adults who have desirable lives. For instance, as part of her job in the fashion industry, Jane takes work trips to Los Angeles, London, and Paris, without her brother's knowledge. By contrast, the teens of *Bunheads* live relatively grounded lives. Other differences between the series include setting, as *Bunheads* relies on a certain amount small town charm, as quirky characters populate Paradise, while *Jane by Design* draws its world around the metropolitan allure of New York City.

Taken together, *Bunheads* and *Jane by Design* highlight the contradictions and complexity of undertaking and representing particular forms of creative feminine labor. In *Bunheads,* this labor takes the form of ballet dancing, and in *Jane by Design* it is represented through the designing and creating of fashion. Both series depict kinds of creative labor which have been traditionally understood as feminine and thus often not treated as "serious" work. Through their negotiation of creative labor, this essay argues that *Bunheads* and *Jane by Design* are part of a new wave of feminist-inflected U.S. scripted television series that are negotiating feminist and postfeminist issues and rearticulating them for and within a postfeminist culture. They are not necessarily "feminist" series, but they are part of a feminist turn in U.S. scripted television more generally, and on ABC Family/Freeform specifically. This is particularly important as most ABC Family/Freeform programs are aimed at young women.[4] *Bunheads* and *Jane by Design* are best understood as in dialogue with the recent resurgence of feminism in Anglophone popular culture, often identified as a fourth wave of feminism.

"Blossoming of new feminisms"

Both feminism and postfeminism are central to understanding the screen culture in which *Bunheads* and *Jane by Design* are operating, better understood as postfeminist media culture. Rosalind Gill coined the term as a way of conceptualizing how in Western popular culture women rather than men are constituted as ideal neoliberal subjects.[5] In media studies, neoliberalism is understood as a set of economic, social, and cultural conditions that shape Western society and reinforce the value of the free market and capitalism.[6] U.S. primetime scripted television from the 1990s to present functions within and contributes to a postfeminist media culture. Postfeminism is a highly contested term, and as such there are various competing definitions and conceptualizations.[7] In her book *Postfemininities in Popular Culture*, Stephanie Genz argues that attempting to fix the meaning of postfeminism is "futile and misguided" given that each articulation of postfeminism is in itself "a definitional act that (re)constructs the meaning of feminism and its own relation to it."[8] Postfeminism is always in conversation with popular culture and constructed through popular culture. For feminist theorist Angela McRobbie, postfeminist culture promotes "a highly conservative mode of feminine empowerment," which has arisen in place of feminism, and is defined by women's newfound sexual freedoms.[9]

Feminist criticism of postfeminism largely focuses on screen texts from the late 1990s and early 2000s, such as *Bridget Jones's Diary* (2001), *Ally McBeal* (FOX, 1997–2002), and *Sex and the City* (HBO, 1998–2004).[10] This work examines the decline of explicitly feminist content, as well as the rise of overtly anti-feminist content. Much of the academic work on *Sex and the City* has characterized the series as postfeminist, due to its perceived anti-feminist but pro-woman sensibility.[11] For example, *Sex and the City* emphasizes empowerment through bodily femininity and "work on the self" in the form of fashion and grooming. In her article "Notes on the Perfect: Competitive Femininity in Neoliberal Times," McRobbie examines how contemporary postfeminist discourses operate in a popular culture in which feminism has made a comeback.[12] In a revision of the prognosis that she laid out in *The Aftermath of Feminism*, McRobbie considers the "blossoming of new feminisms across so many different locations," and "an eventual explosion" of postfeminism's stranglehold on contemporary femininity.[13] McRobbie argues that while women's engagement with feminist activism has increased, there is simultaneously an amplified emphasis on bodily representations of "perfect" femininity and punishment for those who do meet standards of perfection.[14]

While McRobbie does not use the term "fourth wave" to describe this blossoming of new feminisms, other popular culture commentators and feminist critics have proposed that we are currently experiencing a "fourth wave of feminism."[15] The idea of a fourth wave arose in the 1990s; however, it did not really begin to take shape until the early to mid–2000s.[16] One of the earliest academic theorizations of fourth-wave feminism is E. Ann Kaplan's "Feminist Futures: Trauma, the Post-9/11 World and a Fourth Feminism?"[17] Kaplan's discussion of fourth-wave feminism or a fourth feminism is couched in terms of two central concerns. Firstly, her consideration of a possible fourth feminism is framed by the impact of 9/11 on U.S. institutions and women in the so-called "era of terror."[18] Secondly, she anticipates the challenges that will be faced by future feminisms in relation to the legacy of second- and third-wave feminisms.[19] For Kaplan, fourth-wave feminism is best understood as in opposition to earlier iterations of American feminism, which were overtly white, liberal, and middle-class.[20]

The feminisms of *Bunheads* and *Jane by Design* are not clearly second wave, third wave, or postfeminist. Unlike *The Good Wife* (CBS, 2009–2016), which overtly invokes second-wave feminist issues like women in the workplace, or *Orange Is the New Black* (Netflix, 2013–present), which speaks to third-wave feminist concerns about Black and queer women, *Bunheads* and *Jane by Design* engage with contemporary popular feminisms that are not explicitly tied to earlier eras. *Bunheads* and *Jane by Design* both reject and renegotiate many (but not all) tenets of postfeminist culture, and as such they are part of a resurgence of feminism in Anglophone popular culture that has emerged in the last decade and has been discussed by McRobbie and Gill.[21]

While feminist politics on U.S. scripted television takes many forms, in *Bunheads* and *Jane by Design* feminist politics can be seen in how each series depicts the difficulties of engaging in and performing feminine labor in a *post*-postfeminist moment. For example, the characters of *Bunheads* manage their bodies through exercise, stretching, and eating to be "perfect" ballet dancers. In *Jane By Design,* the titular Jane bears the financial and emotional burden of living in a parent-less household, as she struggles to "fit in" both at school and work. In order to understand how *Bunheads* and *Jane by Design* articulate their feminist and postfeminist politics, we need to examine how they situate themselves within various lineages of feminist and feminine screen culture.

Gender as a Classificatory Strategy

From the 1950s, American scripted television was considered a feminized medium associated with women in the domestic space.[22] However, in the post-network era which is characterized by the expansion cable channels and subscription video-on-demand services, the dominant narratives around U.S. scripted television emphasize the idea that it has been masculinized.[23] As Charlotte Brunsdon argues, "instead of being associated with housebound women, this new television is young, smart, and on the move, downloaded or purchased to watch at will."[24] Yet the gendering of U.S. scripted television and television history runs much deeper than its reception. Television scholar Lynne Joyrich argues that "gender is not simply a potential *subject matter* for television—it is a classificatory strategy, a structuring system, a very significant *matter for subjects* constituted through its terms of enunciation and address."[25] Gender informs both the creation and reception of television; this is evident in how *Bunheads* and *Jane by Design* depict creative labor.

Media studies scholar Aniko Imre argues that "television as an object has been gendered from the start," referring to the idea of scripted television drama as a feminine mode of storytelling, which is consumed in the domestic space and used to advertise products to women.[26] Therefore, television's feminine roots are often obscured in the era of high concept, high gloss male-centric "quality" television, such as *Mad Men* (AMC, 2007–2015), *Breaking Bad* (AMC, 2008–2013), and *Game of Thrones* (HBO, 2011–present).[27] On ABC Family/Freeform, however, the feminine pedigree of television is celebrated. Both *Bunheads* and *Jane by Design* appreciate and embrace feminist and feminine screen history, drawing on different lineages in distinct ways. *Bunheads,* for instance, references key feminist texts and figures through dialogue. Throughout its eighteen episodes, different characters mention crucial feminist series such as *I Love Lucy* (CBS, 1951–1957), *The Carol Burnett Show* (CBS, 1967–1978), *The Mary Tyler Moore Show* (CBS, 1970–1977), and the film *A League of Their Own* (1992). Various references to feminist figures, such as Sylvia Plath and Hillary Rodham Clinton are also included. The whip-smart fast-paced dialogue at times creates the illusion of a cohesive feminist politics, but this does not consistently translate to characterization.

While *Bunheads* uses dialogue, *Jane by Design* employs imagery and costuming that recall female-centric romantic comedy films and television series, such as *Working Girl* (1988), *The Devil Wears Prada* (2006), and *Sex and the City*. In the first shot of the series we see curly-haired Jane strutting in heels down a New York City street in a pink tutu. This image, paired with the ironic voice over, clearly invokes the series *Sex and the City* and its protagonist Carrie Bradshaw (Sarah Jessica Parker). *Jane by Design* is aimed at a media-literate audience who understand this reference intuitively, and as such it places itself in a lineage with the iconic (although more adult) television romantic comedy. Other visual homages include the final shot of the second episode, where the traffic clears to reveal Billy leaning against his car, replicating a recognizable shot from the end of *Sixteen Candles* (1984).[28] In a reference to another Molly Ringwald classic, *Pretty in Pink* (1986), Jane redesigns her mother's old pink party dress for a school dance.[29] Furthermore, Jane's faux-hawked punk(ish) best friend Billy has a particular resonance with Andie's (Molly Ringwald) best friend in *Pretty in Pink*, the now-iconic Ducky (Jon Cryer).[30] The Jane/Billy dynamic is very similar to Andie/Ducky's friendship, as they are platonic childhood friends from working class backgrounds who are outcasts in their high schools. While *Sixteen Candles* and *Pretty in Pink* are not feminist classics, they are delightfully feminine in their approach and perspective. They are also classics of the teen pic genre, which is the filmic counterpart to television's teen soap opera.[31]

In addition to visual references to television series and films that celebrate femininity, *Jane by Design* also uses guest appearances from fashion industry professionals to celebrate a particular brand of female empowerment. Notable guest appearances include *Sex and the City* costume designer Patricia Field, *Marie Claire* creative director Nina Garcia, fashion designers Nanette Lepore and Betsey Johnson, Jimmy Choo co-founder Tamara Mellon, and former fashion model Paulina Porizkova. The inclusion of these well-known figures from the New York fashion world allows Jane to operate in a space dominated by powerful, fashionable women. The visible presence of these women emphasizes the absence of Donovan Decker. In doing so, however, it promotes a very narrow version of female empowerment, which is white, thin, and heterosexually desirable, as does the casting of Andie Mac-Dowell as Jane's boss Gray Chandler Murray, the creative director of the fictional fashion house Donovan Decker. MacDowell had a very successful film career in the 1990s, appearing in comedies such as *Green Card* (1990), *Groundhog Day* (1993), and *Four Weddings and a Funeral* (1994). Her presence invokes an earlier era of film romantic comedies, and the kinds of character she was known for—strong, powerful, and funny.

Both *Bunheads* and *Jane by Design* embrace femininity and the history of feminine screen culture. This, however, does not necessarily make the series feminist in point of view or politics. Femininity is celebrated in both series, but ultimately it comes at a cost. For instance, *Bunheads* creates a binary between femininity and practicality. Both Michelle and Fanny are former world-class ballet dancers who are seemingly incapable of running a business. In the world of *Bunheads,* characters who are "good" at femininity are almost always "bad" at practical things. This is reinforced through sisters Millie (Liza Weil) and Truly Stone (Stacey Oristano). The loving, warm, and emotionally fragile Truly is bad at business, but she makes adorable dresses, while the cold, demanding, and emotionally manipulative Millie is a small-town mogul. An investment in and celebration of femininity is not a reliable indicator of a series' feminist sensibility.

Good Labor vs. Bad Labor in Bunheads

Lamar Damon and Amy Sherman-Palladino co-created *Bunheads*, and while Damon is credited as the series' co-creator, Sherman-Palladino authored the pilot script and acted as showrunner and executive producer. Sherman-Palladino is well known for creating cult classic series *Gilmore Girls,* and *Bunheads* shares considerable tonal, aesthetic, and thematic similarities with the earlier series. In many ways *Bunheads* is a spiritual sister to *Gilmore Girls* thanks to the rapid-fire dialogue, music by Sam Phillips, and similar reoccurring guest stars, such as Rose Abdoo, Sean Gunn, and Todd Lowe, who play quirky townspeople in both series. During her writing career Sherman-Palladino has developed her own trademarks, one of which is quirky women who talk fast and like coffee. Her female leads often simultaneously embrace and reject different aspects of traditional femininity. For instance, both *Gilmore Girls'* Lorelai Gilmore (Lauren Graham) and *Bunheads'* Michelle embrace bodily femininity, but they do not cook or undertake other forms of traditionally feminine domestic labor. Michelle, like Lorelai, is a traditionally attractive white woman, and she uses this to get things, but she is not defined by it—meaning her self-worth is not tied to her heterosexual desirability. Despite their many similarities, *Bunheads* operates in a distinctly different world to *Gilmore Girls*, not just geographically; *Gilmore Girls* is set in Connecticut and *Bunheads* in California. *Bunheads'* world is shaped by, engages with, and incorporates dance as a mode of creative expression and labor.

The first episode of *Bunheads* opens with Michelle—a Las Vegas showgirl—dancing as part of chorus of dancers wearing large feathered headdresses and bedazzled bikini-like costumes.[32] Towards the end of the number the "real dancers" move to the back of the stage and the topless showgirls enter to cheers from the audience. Here we are introduced to Michelle and her best friend and fellow dancer Talia (Angelina McCoy), as well as the class politics of the dance world. While the diegetic audience's attention is on the topless dancers, the camera focuses on Michelle and Talia as they dance at the back of the stage and discuss how they get paid less than the topless dancers for doing more work. Michelle asserts: "It's a great message we are sending to the girls of America. Hey girls, forget about actually learning how to dance, just take your top off and stand there."[33] There is a clear division between the dancers who are willing to take their clothes off and those who are not, and the series clearly aligns itself with those who choose not to. From the very first scene of the series there is a clear stratification of feminine labor that deems a particular kind of dancing high-class and other kinds low-class. From its opening scene, *Bunheads*' feminist politics are messy at best.

Bunheads goes to great lengths to simultaneously distinguish Michelle as *not* a sex worker, while associating her with sex work. In the early episodes, in particular in the pilot episode, Michelle is often located in relation to sex work. For example, Michelle lives next door to a supposed sex worker who is the physical opposite of her. Michelle is tall, brunette, and thin, while her neighbor is short, blond, and fat. Later, Michelle's behavior also works to distinguish her from a sex worker, when, after a bad audition, Michelle agrees to go out for dinner with her longtime admirer Hubbell (Alan Ruck). During the dinner Michelle treats him with distain, gets very drunk, and agrees to marry him and move to Paradise, California. Despite her open hostility, Hubbell is kind and patient with Michelle. How Hubbell treats Michelle is contrasted with how the entertainment industry treats and exploits women. He sees what the industry and Las Vegas cannot: her talent, beauty, and potential.

Upon Michelle and Hubbell's move to Paradise, she is not warmly welcomed to the sleepy beachside community, where a running joke develops throughout the series in which characters identify her previous occupation as different variations on sex work. She is called a stripper, hooker, *Playboy* playmate, and pole dancer. This joke relies on the knowledge that Michelle is *not* a sex worker, and on the assumption that being a sex worker is humorous. Sex work is not seen as work by the series, and there is definite derision of sex workers that pervades the series. This is a seemingly conservative and anti-sex stance for a series that emphasizes bodily labor in the form of dancing. In the world of *Bunheads*, ballet dancing is "good," while sex work is "bad" work. This way of thinking is continued later in the series when the central teen characters ridicule the kick height of the cheerleaders, making the clear distinction between ballet as "real" dancing and cheerleading as "bad" dancing.[34]

The first episode of *Bunheads*, surprisingly, ends with Michelle finding out her new husband Hubbell has been killed in a car accident. By killing off Hubbell at the end of the first episode, the series' narrative impulse is dramatically altered. *Bunheads* is transformed from a damsel narrative, whereby Michelle needs Hubbell to save her from her unhappy life, to a story about finding independence and a home. The series is no longer about Michelle being saved by a man (namely Hubbell), but it becomes about her finding her purpose and building a new life for herself in Paradise. The primary relationship of the series, therefore, is not between Michelle and Hubbell, but between Michelle and Hubbell's mother Fanny. When Hubbell dies, the house and dance studio is left in the will to Michelle; therefore, Michelle and Fanny must live and work together. As such, *Bunheads*, constructs a diegesis where multiple generations of women live, work, and dance together.

Bunheads is not a series that just incorporates dance or is about ballet dancers, but it is a series that wholly celebrates dance as a form of emotional, thematic, and personal expression. The audience learns as much about the characters from how they dance as from what they say or how they act. The series features both unfinished dance numbers in progress and fully choreographed and costumed dances in different styles, from modern contemporary to classical ballet. The potential of the dance numbers to convey emotional and thematic meaning is most clearly seen when Sasha dances to "Istanbul (not Constantinople)" by They Might Be Giants.[35] In this episode, Sasha is forced to confront the demise of her parents' marriage, which she has been in denial about for most of the series. The dance number begins with an angry, heavily eye-linered Sasha flanked by two other unnamed dancers taking up the frame. All are wearing black dance tights with black turtlenecks, as they stare down the camera and the audience. The dance works to express Sasha's anger towards her parents and general teenage angst. Each episode includes from one to three dance numbers that at times (although not always) work to advance the narrative. The dance numbers have an emotional and thematic resonance with the larger story and characters' journeys, but they also operate as a celebration of dancing as a feminine art form. I contend that the dance numbers function as at a metaphoric level to celebrate television as a feminine art form.

Considerable screen time is dedicated to scenes where the characters dance. These long, unedited dance numbers are mostly filmed in the style of classical Hollywood musicals, such as *42nd Street* (1933), *Top Hat* (1935), and

Singin' in the Rain (1952), in one take and with few close-ups or cut-ins. The camera shows the actors' full body movements as a way of appreciating the dancing and to assert that body doubles are not being used. The camera becomes part of the choreography, pushing forward and pulling back in time with the music. The casting of real dancers reinforces the authenticity of these dance numbers, in particular the casting of Broadway darling Sutton Foster, whose Tony Award–winning singing and dancing underwrites Michelle's talent. Foster/Michelle's talent operates as a kind of insurance policy that allows Michelle to remain a likeable and sympathetic protagonist, even though she struggles as a teacher and businesswoman. For the series to revel in and celebrate dancing, the audience must believe in the dancer's ability. This can also be seen in the casting of musical theatre veteran Kelly Bishop as former professional ballet dancer Fanny.[36]

The bodily labor of ballet, such as the stretching and repetition of movements, is so ubiquitous that it becomes part of the mise-en-scène of *Bunheads*. While their work is punishing on the young girls' bodies, there is no guarantee that their hard work will lead to success in the dance world. For Ginny and Boo, no amount of work is going to make their bodies (both of which are too curvy) suitable for professional ballet, a representation of the reality of the ballet world. This emphasis on bodily labor and the work of dancing is also played out through a gross feet competition.[37] The titular bunheads take photographs of their blistered, scarred, deformed feet as part of a competition with other local ballet schools. The dancers with the grossest feet are considered the best dancers. Their feet represent their labor and talent. The vulgarity of their bodies is celebrated as evidence of their hard work, and it operates in contrast to the stereotypical view of ballet dancers being delicate swan-like creatures who create beautiful art, seemingly without effort or work.

The labor of dancing is made visible by the series, as are the characters' bodies. The shooting style of the dance numbers ensures that the dancers' bodies are seen in their entirety. *Bunheads* not only allows, but also encourages the characters to take up space within the frame. This translates outside the dance studio, as Sasha, Boo, Ginny, and Melanie take up space diegetically at the cinema, at school, and at restaurants. They are unapologetic in their interactions with the world, such as how they request free fried food at a local restaurant or manipulate movie-goers to get four seats together at a sold-out film. *Bunheads* engages critically with postfeminist notions of constructing the female body, which suggest that women's bodies are simultaneously powerful and "always already unruly and requiring constant monitoring, surveillance, discipline and remodeling."[38] In *Bunheads*, artifice is rejected in favor of visible labor. While the series' depictions of dancing as feminine labor can be seen as in dialogue with larger fourth-wave feminist concerns, its repeated disparaging of sex work complicates a reading of *Bunheads* as "feminist."

Creative Labor as Postfeminist Affirmation in Jane by Design

In the same way that *Bunheads* embraces dancing as a form of physical, emotional, and thematic expression, *Jane by Design* uses fashion and textiles as a mode of self-fashioning. Within the series, fashion is used to mark both Jane and Billy as "other" at their high school, but in the professional world Jane's "look" gives her access to economic and professional freedom. Fashion is situated by the series as a self-discovery tool, rather than a branding tool. Jane expands on young adults and fashion trends, "Being a teenager is about discovering who you are, not just changing who you are to fit some ridiculous mold."[39]

This idea of fashion's potential for self-discovery and empowerment in reinforced by the extra-diegetic opening credit sequence that opens every episode. The credit sequence depicts Jane entering her closet, separating the clothes, and walking through to a New York cobblestone street where a fashion shoot is taking place. She takes a belt off the rack and places in on the model to an approving look from Gray. Jane's closet transports her to her fantasy, where she is empowered to live her dream. The final two shots of the credit sequence show Jane walking towards the camera looking strong and happy, followed by a long-shot of Jane walking past a boutique with show's title in large font on the window. Over these final shots, a high-pitched but sweet female voice sings, "life is a work of art." This is a central theme of the series, which invests in fashion's transformative potential.

While the characters of *Bunheads* do not experience any dire financial issues, in *Jane by Design* many of the characters, including Jane, Billy, and Ben, make choices based on economic need. Therefore, there is a layer of class awareness that pervades the series' approach to labor. Jane is from a lower class background, her mother is absent, and after her father's death, her older brother gave up a promising baseball career to come home and take care of her. The series' instigating event—Jane being mistaken for an adult and offered an executive assistant job instead of an internship—is premised on more than her desire to pursue her dream job in fashion; there is a financial demand. Jane's brother Ben is having trouble finding a job, and he has fallen far behind in the bills. Jane's new job provides financial stability for her and Ben.

While *Bunheads* goes to great pains to make the labor of dancing visible, *Jane by Design* often operates around the logic of divine inspiration. There are moments where creative labor is shown and seen, but they are often accompanied sudden instances of inspiration that seems to strike out of nowhere. When the work of making fashion is shown, it almost always in service of a particular deadline or presentation, such as when Jane puts together a preview of Donovan Decker's upcoming jewelry line for investors.[40] Like much of the creative labor in the series, the work of putting together the presentation is shown through montage. Jane's work, and the work of everyone at Donovan Decker, is framed in terms of success or failure, which is generally judged by external sources—investors, clients, and the market.

The labor of fashion is highlighted in the series' many makeover plots. Fashion is framed as enabling characters to access social acceptance or professional success. The makeover plot is a staple of postfeminist media culture, teen pics, and romantic comedies, such as *Pretty Woman* (1990), *Clueless* (1995), and *The Princess Diaries* (2001). In *Jane by Design*, the makeover plot is also the solution to many of the characters' personal problems. For instance, when Billy starts dating popular girl Lulu (Meagan Tandy), he changes his "look" to gain approval from her father and friends. Billy transforms himself from punk-lite (faux-hawk, suspenders, leather jackets, studded cuffs and belts) to preppy (polo shirts, chinos, pastels, button-down shirts, and bangs). Another example is when high school guidance counselor Rita (Smith Cho) tries to get Ben's attention. She changes out of her conservative button down blouse and skirts into tight-fitting, sparkly, low-cut clothes more suitable for a nightclub than a high school. Her new outfit is paired with big bouncy hair and a full face of makeup. With both of these transformations, we do not see the work, just the result. In *Jane by Design* the makeover plot is used to highlight the transformative potential of fashion, and by proxy consumption and capitalism.

Ultimately all of the creative labor undertaken in *Jane by Design* is subsumed under the name of a never-seen male figure—Donovan Decker. For example, when the company is tasked with designing a wedding dress for "American royalty" Charlotte Whitmore (Autumn Reeser), it is the mysterious Donovan Decker who gets the credit for Jane's design.[41] Of course this is standard practice in corporations with a figurehead, but it is also another way that women's work is rendered invisible. In *Jane by Design*, it is primarily women—Jane, Gray, and associate designer India (India de Beaufort)—who we see doing the bulk of the creative labor. Yet it is accepted and acceptable that their work is undertaken in the name of Donovan Decker, who is a placeholder for the many (presumably white) male fashion designers who dominate the fashion industry. While the series acknowledges and appreciates Jane's work, it is not "seen" by the world inside the series.

Much of Jane's professional success is positioned not as the result of her hard work, but rather thanks to external intervention. Many *Jane by Design* episodes revolve around Jane encountering and overcoming a work problem with the help of Billy. For instance, when a trunk show at Gray's house turns into a raging party, Billy helps Jane shut it down and then clean up before anyone finds out,[42] or when Jane's colleague India shows up at their high school and threatens to uncover Jane's double life, Billy creates a distraction.[43] Billy repeatedly saves Jane's job and professional standing, and he enables her to live a double life. Throughout the series, it seems that Jane is incapable of saving herself, for although she has many of the markers of Murphy's independent ABC Family heroine, she is content to work within the patriarchy.[44] Ultimately, *Jane by Design* operates as an affirmation of postfeminist culture, rather than as a challenge to it.

Conclusion

Bunheads and *Jane by Design* show how subversive feminist politics can and do operate in unlikely places, such as in family-friendly programing. *Bunheads* is doing at a series level what the characters within each series cannot or will not do—they are recognizing feminine creative labor. *Jane by Design*, however, struggles to make the work, as well as the results of that work, visible. The feminist politics of *Bunheads* and *Jane by Design* are as much extra-textual as they are textual, insofar as much of their feminist resonance extends from references to earlier feminist films and television series and the invoking of powerful female figures. At the same time, the feminist politics of both series are contradictory, neoliberal, and flawed. While *Bunheads* makes visible (both to the audience and within the diegesis) the labor of dance, *Jane by Design* makes the work of fashion somewhat visible to the audience, but it remains invisible within the show's world. *Jane by Design* often operates as an affirmation of postfeminism, rather than as a challenge to it, as it uncritically employs postfeminist tropes such as the makeover plot and empowerment through fashion and grooming. While *Bunheads* at times privileges the individual over the collective, it does depict the titular dancers working together on various occasions for a common good, such as putting together a dance number to help Fanny grieve the death of Hubbell.[45] *Bunheads* and *Jane by Design* are part of a larger conversation about popular feminism on U.S. scripted television, and while neither series fulfills its feminist potential or translates its feminist intentions fully, they are both negotiating the complexity of postfeminist culture and the contemporary feminist moment. As such, they are both contributing to and shaping this fourth wave moment.

NOTES

1. Amanda R. Keeler, "Branding the Family Drama: Genre Formations and Critical Perspectives on *Gilmore Girls*," in *Screwball Television: Critical Perspectives on Gilmore Girls*, ed. David Diffrient and David Lavery (Syracuse: Syracuse University Press, 2010), 19–20.

2. For a discussion of the "happy family drama," see Barbara Moore, Marvin R. Bensman and Jim Van Dyke, *Prime-Time Television: A Concise History* (Westport, CT: Praeger, 2006), 187–188.

3. Caryn Murphy, "Secrets and Lies: Gender and Generation in the ABC Family Brand," in *The Millennials on Film and Television: Essays on the Politics of Popular Culture*, ed. Betty Kaklamanidou and Margaret Tally (Jefferson: McFarland, 2014), 26.

4. Michelle Castillo, "ABC Family Doubles Down on Original Programing to Reach Young Women," *AdWeek*, April 14, 2015.

5. Rosalind Gill, *Gender and the Media* (Cambridge: Polity Press, 2007), 249.

6. Nick Couldry, *Why Voice Matters: Culture and Politics After Neoliberalism* (London: Sage, 2010), 4.

7. For surveys of the competing definitions of postfeminism see Rosalind Gill and Christina Scharff, eds., *New Femininities: Postfeminism, Neoliberalism and Subjectivity* (New York: Palgrave Macmillan, 2011) and Sarah Projanksy, *Watching Rape: Film and Television in Postfeminist Culture* (New York: New York University Press, 2001).

8. Stephanie Genz, *Postfemininities in Popular Culture* (Hampshire: Palgrave Macmillan, 2009), 19.

9. Angela McRobbie, *The Aftermath of Feminism: Gender, Culture and Social Change* (London: Sage, 2009), 26–27.

10. See McRobbie, *Aftermath of Feminism*; Yvonne Tasker and Diane Negra, eds., *Interrogating Feminism: Gender and the Politics of Popular Culture* (Durham: Duke University Press, 2007); and Diane Negra, *What a Girl Wants? Fantasizing the Reclamation of Self in Postfeminism* (London: Routledge, 2008).

11. See Diane Negra, "'Quality Postfeminism?': Sex and the Single Girl on HBO," *Genders* 39 (2004), http://www.colorado.edu/gendersarchive1998–2013/2004/04/01/quality-postfeminism-sex-and-single-girl-hbo.

12. Angela McRobbie, "Notes on the Perfect: Competitive Femininity in Neoliberal Times," *Australian Feminist Studies* 30, no. 83 (2015): 4.

13. *Ibid.*, 9.

14. *Ibid.*

15. See Jennifer Baumgardner, *F'em! Goo Goo, Gaga, and Some Thoughts on Balls* (Berkeley: Seal Press, 2011); Kira Cochrane, *All the Rebel Women: The Rise of the Fourth Wave of Feminism* (London: Guardian Books, 2013); and Rosalind Gill, "Post-Postfeminism? New Feminist Visibilities in Postfeminist Times," *Feminist Media Studies* 16, no. 4 (2016): 610–630.

16. For early discussions of what form the fourth wave may take and how it may emerge out of the third wave, see David Golumbia, "Rethinking Philosophy in the Third Wave of Feminism," *Hypatia* 12, no. 3 (1997): 100–115.

17. E. Anne Kaplan, "Feminist Futures: Trauma, the Post 9/11 World and a Fourth Feminism?" *Journal of International Women's Studies* 4, no. 2 (2003).

18. *Ibid.*, 47.

19. *Ibid.*, 55.

20. *Ibid.*, 49.

21. McRobbie, "Notes on the Perfect," 9; Gill, "Post-Postfeminism," 612.

22. Lynn Spigel, *Make Room for TV: Television and the Family in Postwar America* (Chicago: University of Chicago Press, 1992), 73–74.

23. See Michael Z. Newman and Elana Levine, *Legitimating Television: Media Convergence and Cultural Status* (New York: Routledge, 2012), 10–11.

24. Charlotte Brunsdon, "Bingeing on Box-Sets: The National and the Digital in Television Crime Drama," in *Relocating Television: Television in the Digital Context*, ed. Jostien Gripsrud (London: Routledge, 2010), 65–66.

25. Lynne Joyrich, *Re-Viewing Reception: Television, Gender, and Postmodern Culture* (Bloomington: Indiana University Press, 1996), 17. Original emphasis.

26. Aniko Imre, "Gender and Quality Television," *Feminist Media Studies* 9, no.4 (2009): 392.

27. See Newman and Levine, *Legitimating Television*, 12.

28. "The Runway," *Jane by Design*, season 1, episode 2, ABC Family, January 3, 2012.

29. *Ibid.*

30. It is worth noting that Howard Deutch, director of *Pretty in Pink*, also directed an episode of *Jane by Design*, "The Getaway," season 1, episode 9, ABC Family, February 28, 2012.

31. See Timothy Shary, *Teen Movies: American Youth on Screen* (London: Wallflower Press, 2005), 67–72.

32. "Pilot," *Bunheads*, season 1, episode 1, ABC Family, June 11, 2012.

33. *Ibid.*

34. "No One Takes Khaleesi's Dragons," *Bunheads*, season 1, episode 9, ABC Family, August 13, 2012.

35. "Movie Truck," *Bunheads*, season 1, episode 6, ABC Family, July 23, 2012.

36. Notably, Kelly Bishop created the role of Sheila in the original Broadway production of *A Chorus Line,* winning a Tony Award for her performance in 1976.

37. "Inherit the Wind," *Bunheads*, season 1, episode 3. ABC Family, June 25, 2012.

38. Gill, *Gender and the Media*, 255.

39. "The Image Issue," *Jane by Design*, season 1, episode 6, ABC Family, February 7, 2012.

40. "Pilot," *Jane by Design*.

41. "The Wedding Gown," *Jane by Design*, season 1, episode 8, ABC Family, February 21, 2012.

42. "The Birkin," *Jane by Design*, season 1, episode 3, ABC Family, January17, 2012

43. "The Image Issue," *Jane by Design*.

44. Murphy, "Secrets and Lies," 26.

45. "For Fanny," *Bunheads*, season 1, episode 2, ABC Family, June 18, 2012.

How to *Make It or Break It*
Empowering Girls in a Neo-Feminist Era

MADELINE RISLOW *and* ANNE DOTTER

The ABC Family original show *Make It or Break It* (2009–2012) was created and mostly written by Holly Sorensen. While Sorensen is best known for *Make It or Break It* (*MioBi*), her approach to the show was shaped by her previous experience as the assistant to feminist activist Gloria Steinem, and as a senior editor for the film magazine *Premiere*. *MioBi,* which debuted in prime time on June 22, 2009, to 2.5 million viewers, aired for twenty episodes each of its first two seasons. The show was eventually renewed for a final eight episodes of season three, which aired in early to mid-spring 2012, "in time to ignite Olympic fever," in the words of ABC Family president Michael Riley.[1] Beyond a setup for that summer's London Olympics, *MioBi* is a teen show in very good company on ABC Family, the network that also created *Kyle XY* (2006–2009), *Greek* (2007–2011), *Switched at Birth* (2011–2017), among many others, all demonstrating that long form television allows for more depth and different explorations of character developments than the more constraining feature film.

During its three-year run, *MioBi* presented girls with both physically and emotionally empowering messages. The storyline follows Payson Keeler (Ayla Kell), Kaylie Cruz (Josie Loren), Lauren Tanner (Cassandra Scerbo), and Emily Kmetko (Chelsea Hobbs),[2] four American teenage girls from diverse socioeconomic backgrounds as they train together at Rocky Mountain Gymnastics Training Center (aka "The Rock," an elite fictional facility located in Boulder, Colorado). The girls compete throughout the series run for a spot on the USA gymnastics team for the 2012 London Olympics. Instead of glossing over challenges and differences, as is the norm in girly media,[3] producers at ABC Family elected to use the narrative as a conduit to put front and center culturally-sanctioned and encouraged masochistic behaviors (eating disorders, serious bodily injuries) and other issues such as socio-economic differences, friendship and jealousy, blackmail, romance, sexuality, and unplanned pregnancy. This is achieved through the distinct personalities and character development of the four girls explored throughout the series, as well as through a direct narrative address of the issues at stake.

Despite its provocative take on teen girls' condoned behaviors, Jessica Bendinger's 2006 feature film *Stick It,* also set in the world of elite women's gymnastics, follows a classic Cinderella narrative framework and thereby provides an ideal object of comparison to clearly establish the many ways in which *MioBi* stands out as a challenge to mainstream neo-feminist media. Indeed, *Stick It* follows 17-year-old Haley Graham (Missy Peregrym) as she is forced to return to competitive gymnastics after a run-in with the law.[4] Akin to Cinderella, Haley holds the promise of a bright future (she was an elite gymnast); when the film starts, however, she is an outcast, just like Cinderella at the onset of her own narrative. Cinderella narratives tend to follow a three-part arc including a lowly beginning, a makeover and a coming-out moment, which affirms the lead character as a princess as well as affirming heteronormative expectations. Neither *Stick It* nor *MioBi* strays from the affirmation of the heterosexual norm, as none of the central female characters ever seem to question their heterosexual orientation. *Stick It*'s narrative arc focuses on restoring Haley to her leadership position in USA elite gymnastics: from her makeover (the intense month of physical training required to get her back in competitive shape), to her coming-out ball (Nationals, when she gets to settle her differences both with fellow competitors and with the entire gymnastics world including her mother, as well as her former coach and the judges), *Stick It* follows a standard Cinderella storyline in which Haley's white body is placed front and center as the primary gauge of her successful makeover.

In *MioBi*, white bodies are also front and center at all times: young women's physiques are shown in leotards throughout, thereby revealing much of their bodies to the viewer's gaze. But these bodies, *MioBi* tells us, are special. The opening credits focus on the performances of their routines and the diffusion of chalk dust each time the gymnasts hit an element, celebrating their bodies' abilities and highlighting the danger of each and every one of the tricks they perform rather than objectifying the women. Payson's statement "Our bodies are everything we have. Everything we work for"[5] is made in response to comments about sex and the empowering nature of it. For Payson, it is control over her body that is empowering, not being the object of someone else's enjoyment. The line is drawn: the four gymnasts we follow over the course of three seasons and 48 episodes are exceptional in many more ways than their counterparts in shows and films made about, and generally considered for, teen girls. They are in control of their bodies, because they have to be, not to please, nor to respond to social norms.

We begin this essay with a brief discussion of the contribution we are making to the scholarship at the intersection of teen media, feminist media, and sports films scholarship. Then we concern ourselves with narrative structure and aesthetic decisions before moving to the unique contributions *MioBi* makes, most notoriously the focus on such issues as privilege and wealth, as well as body image and eating disorders. We argue that *MioBi* opens new avenues for the representation of young white feminine identities, alternatives not grounded in consumerist culture or the aspirations written in narratives driven by a Cinderella model.

Despite its centering of whiteness, *MioBi* challenges representations of teen gendered norms. Indeed, producers for the show do not shy away from difficult conversations such as eating disorders or the challenges faced by young women and their body image during their teen years. *MioBi*'s selection of societal issues is narrow: it glosses over problems related to disability, race, and sexuality by paying lip service to them. The issue of physical disabilities, for example, is present in two marginal ways in *MioBi*: Emily's little brother uses a wheelchair,[6] and temporarily when Payson, after a serious back injury, envisions a life without the ability to walk, let alone be a gymnast.[7] Despite the common racial diversity of USA gymnastics, a black lead gymnast is not introduced until the final shortened season.[8] None of the four lead characters are constructed such that they would challenge racial or sexual identity norms: all four are constructed as cis-gendered, white, able-bodied girls inclined toward heteronormal relationships, and generally, the narrative seems to idealize the nuclear family as a source of stability for young adults. The show favors a focus on socioeconomic differences to the detriment of race or sexuality, thereby engaging directly with neo-feminism.

Scholarly Intersections

Feminist media scholars such as Diane Negra, Yvonne Tasker, and Angela McRobbie explore the meaning of a cultural era, from the 1990s on, often referred to as postfeminist across their scholarship. This concept is understood in a number of conflicting ways, the consequence of having emerged in popular culture and not scholarly analysis.[9] They recognize the ambivalence of the term: it ushered in a new era, one taking distance from the second wave U.S. American feminists' world view in that it rejects the simplistic binary thinking and essentialism that undergirds the movement. Yet, it also embraces neoliberal values insomuch that individual freedom to choose is affirmed within the constraints of consumer behaviors. Tasker and Negra go even further, warning that "postfeminism signals more than a simple evolutionary process whereby aspects of feminism have been incorporated into popular culture—and thereby naturalized as popular feminism. It also simultaneously involves an 'othering' of feminism (even as women are more centered), its construction as extreme, difficult and unpleasurable."[10] Between object of male desire and independent agent, the scholarship on women in film and media, and more narrowly teen girls' media, has focused on the varying degrees of agency afforded women on the screen since its inception. Feminist critiques such as those offered by Tasker and Negra point out that within a postfeminist cultural zeitgeist, individual women's agency is framed within neoliberal standards, and so are their representations on the screen.

Further affirming this point, Hilary Radner makes it clear in *Neo-Feminist Cinema* that women's ability to affirm agency of any kind happens within the limits set by neoliberal narratives: "Choice and the development of individual agency [are] the defining tenets of feminine identity [within neofeminist media]—best realized through an engagement with consumer culture in which the woman is encouraged to achieve self-fulfillment by purchasing, adorning or surrounding herself with the goods that this culture can offer."[11] Contrary to Tasker, Negra, and McRobbie, Radner repositions neo-feminism within the history of U.S. American feminisms, starting with Helen Gurley Brown's reification of the 1960s single working girl as the empowering parallel to Betty Friedan's *The Feminine Mystique*.[12] Individual freedom of choice as expressed by consumer behavior (fashion, makeup) and pleasure (sexual first and foremost and consumption a close second) is the marker of this empowerment. However, Radner highlights the 1960s sexual revolution's limitations: availability of the female body and its objectification in popular media (such as the Bond girl) does not in effect alter the power dynamic between genders, nor does it affect the social balance. Women still aspire to marry (even if later in life), and even when they do not, they are still engaging in the objectification of their bodies for the sole pleasure of men.[13]

Without engaging in the depth of the feminist conversation, film historian Celestino Deleyto concerned himself with the increasing complexity of media, and the plurality of messages inscribed within. In his book *The Secret Life of Romantic Comedy*, Deleyto argues that the hybridity of genres the romantic comedy tradition has borrowed from since the silent era is only increasing in twenty-first-century productions, thus leading to more layered takeaways, moving away from the simple reaffirmation of conservative values (a common critique of rom-com and other so-called girly media). In fact, he suggests that in the majority of such films, female friendships and the support women can give each other takes precedence over any other message the film might include.[14] The TV show under consideration in this essay is characterized by the hybridity described by Deleyto while resisting the neo-feminist model offered by Radner and critiqued by Tasker, Negra, McRobbie, and others. Both *Stick It* and *MioBi* encompass traits of the comedy, the romance, and the drama, thereby engaging with aspects of neo-feminist narrative strategies. *MioBi* in particular resists the limits imposed by such a framework. Both productions also importantly belong to the genre of sports media, a genre that is more often than not the prerogative of men.[15]

In "Blood, Sweat and Tears: Women, Sport and Hollywood," Katharina Lindner argues that sports films typically lack female protagonists, and when they are present, they are usually portrayed in isolation—no female group dynamic.[16] Furthermore, the female body is typically "marginalized, stigmatized, and/or sexualized" rather than celebrated for the countless hours of hard work required to create a physically powerful version of her body, capable of performing remarkable feats.[17] The same also can be said of a vast majority of portrayals of women in sports on television. *MioBi*'s girls confront common teen girl issues, but they distinguish themselves by never wavering on the empowerment of their athletic identity. In fact, time and again throughout the series we are reminded (by their coaches, parents, the National Gymnastics Organization, the girls themselves—especially Payson) that these girls are not just athletes but *elite* athletes.

Beyond their abilities, Lindner explores the limited character options for women in sports films, further positing that "what is additionally remarkable in this postfeminist context is that the female sporting protagonists tend to be rather isolated figures, with the exception of the varyingly strong bonds to 'supporting' male characters, often their trainer or coach. In particular, they tend to be cut off from other female characters and there is often no hint of a sense of community, female bonding or the opportunities for shared achievement and fun that sports provide."[18] She notes that *Stick It* is one of a handful of notable exception to this. *MioBi* pushes this logic even more, thereby falling straight in line with Deleyto's suggestion that the main takeaway for more recent film and media productions for women (and by extension teen girls) is the focus on the value of women's bonds and the empowerment they will find there. *Stick It* focuses on one tomboyish girl who spends much of the movie insistent upon doing everything herself and only later bonds with the other girls in defiance of judging biases at the USA Gymnastics National Championships. The television show format allows for *MioBi* to create a much more nuanced exploration of the teenage girl experience by following four distinct female characters and their complicated and ever-evolving relationships with each other. Characters such as National Committee member (and later team coach) Ellen Beals (Michelle Clunie) are utilized in *MioBi* to suggest a generational shift within the affirmation of power available for women: following the second wave agenda (individuality), Ellen is tireless in her efforts to break the bond of the *MioBi* girls to satisfy her own aspirations. The four lead protagonists, on the other hand, are systematically successful in their endeavors when they bond and support each other, and fail when they are divided, thereby affirming a new feminist order: one that promotes collective good over individual goals.

Narrative Structure: Resisting Cinderella

MioBi does not concern itself with making its lead characters princesses: they are seldom seen shopping, using makeup, or really engaging in stereotypical teen girl activities (a fact that is frequently highlighted by all four girls and their families as they lament the "normal" life they have never had and that they will never have). The aesthetic and narrative strategies employed in *MioBi* differ from *Stick It* in ways that betray more than a simple difference in media. The long form of the television show allows for further character and plot development. Not only is the Cinderella narrative not needed as the choice arc for *MioBi*, it is explicitly resisted: one of the four lead protagonists, a princess in her own way (because of the privileges afforded her by her father's wealth as well as her whiteness), is the main narrative drive throughout the three seasons. From the pilot episode, Lauren drives the narrative by getting her father (Steve Tanner, played by Anthony Starke) to take away the coach from her peers. Therefore, by the end of the first episode, the power dynamic in the group is already set: the center of agency moving this drama along is Lauren. More broadly, such a narrative strategy centers whiteness at the same time as it empowers young women to take ownership of their own narratives.

MioBi does not shy away either from the ambivalence of Lauren's character: her driving force leads her to be full of contradictions exemplary of teen girlhood, but also more generally reminiscent of the bad girl. Here, however, she is neither rewarded, nor condemned, for embracing her agency. The approach to sexuality is the best way to illustrate Lauren's affirmation of her agency and its meanings: young gymnasts are training so hard that their menses are sometimes delayed and their bodies develop slowly, also delaying the possibility for, or interest in, sexual activity. As Ann Chisholm posits, the American media makes it easy for spectators to consume images of female gymnasts. The media constructs stories that present the athletes as sheltered, child-like, and "cute," which aligns with the visual evidence in their small breasts and hips. Together these attributes deemphasize their sexuality.[19] This halted physical development visually highlights the girls' becomingness, an important characteristic of neo-feminist media: they are changing into young women, therefore holding the promise of sexual pleasures. This all-encompassing nature of the sport controlling teen girls even in the most intimate ways makes sex a very important conversation indeed. When Lauren engaged in sex for the first time, the girls' reactions appear to have been didactically structured such that they offer different possible views of femininity. Lauren affirms that it was empowering, insomuch that she is now in control of her full body; the other three are either mystified and envious (Kaylie), or scared of allowing themselves to do something that might allow uncontrolled changes in their bodies (Payson), or completely unimpressed (Emily).[20] These ascetic responses to active sexuality resist the Cinderella narrative as well as the engagement in pleasure normalized in post-feminist media.

Lauren did not just have sex, but she also stole her best friend's boyfriend. In the process she affirms the ambivalence of her character. She is both bad and fearless. Constructing this character as such gives *MioBi* writers the ability to highlight the power of socioeconomic privilege in a fundamental way: without her father's money, and her white identity, Lauren would not be able to wield such power nor to move the plot the way she does. Sex, narratively, becomes the ultimate way in which Lauren embraces her agency, yet it simultaneously objectifies her, thereby fully illustrating Radner's read of neo-feminist media. While sexual availability and objectification are the locus of a complex relationship between agency and lack thereof in neo-feminist media, such narratives always affirm a right identity. For example, *Stick It* puts Haley front and center as the princess, minimizing all other possible feminine identities by making secondary protagonists fall in the backdrop. On the other hand, because Lauren's behavior is not the norm in *MioBi*, but one elected behavior, questioned by the other three protagonists, the audience is empowered to choose a fitting feminine identity for themselves.

Societal Issues: Privilege

Neo-feminist media erases privileges either by making no character wanting in any way (middle class or beyond), by centering whiteness (by the absence of different races), or by selecting a narrative framework (the makeover or Cinderella narrative) that leads to the affirmation or restoration of the princess to her privileges.[21] This is the case for *Stick It*'s Haley, whose divorced parents have invested in her gymnastics career her entire life, leading her to be an elite gymnast competing at the international level in her mid-teens, yet turning her back to everything she had ever worked for, and losing it all. The narrative arc *Stick It* follows is that of the restoration of Haley's place and privilege. Haley, furthermore, is a composite of three of the four lead characters in *MioBi*: most obviously, her background positions her in Kaylie and Lauren's shoes, as a child raised in privilege and having enjoyed the luxury of elite training from an early age on. However, within the film's diegesis, her rebellious nature has echoes of Emily, what with the alternative way to dress, as well as the association with underground cultures such as BMX /street life, or punk-rock music (for example, Max Morgan's "Ya Better Believe" theme song for Emily in *MioBi*, and t-shirts of the Ramones or Motorhead for Haley in *Stick It*). While Haley's rebellion stems from her parents' divorce, for which she blames herself since it is the product of her mother's affair with her coach, Emily's character is constructed this way to further flesh out her outsider status. She, after all, is from the streets.

More importantly, *MioBi* utilizes all four main characters to paint a picture of American society's socioeconomic diversity, but fails to be inclusive of racial, ethnic, and other diversities until the final season. The three levels of family income and social capital constructed in *MioBi* contrast with *Stick It*, where no socioeconomic differences between gymnasts are apparent. *MioBi* makes socioeconomic background one of the important narrative drives and the justification for the four lead characters' identities: Payson, of middle-class origin, is defined by her work ethic and laser-sharp focus, making her the number one gymnast at The Rock when the series begins; Kaylie is more introspective and less confident in her talents but is driven by the legacy of her own parents' successes and failures; Lauren plays the role of the spoiled rich slut on the surface, but we quickly come to understand that her behavior reflects deeper issues; and Emily's determination to reach elite gymnast status, despite her family's poor economic standing and limited opportunities, has resulted in a hardened, fiercely independent spirit. *MioBi* addresses the problem of class straight on, without attempting to normalize Emily and her family, or romanticizing their lower income status.

When the pilot opens, Emily is the new girl at the gym. She was brought in on the basis of her talent thanks to a scholarship. The "playground prodigy" is dressed to show her lower income, but also as a means to position her as a rebel and an underdog.[22] Where her competitors and friends have had the privilege of training at formal gyms and special camps with elite coaches for most of their lives, she has taught herself most of her early moves on playground equipment and in back alleys, before starting to train in gyms at YWCAs. Emily is not made to feel welcome at The Rock, and often shares that she "does not belong."[23] Other than Payson, for whom little counts outside of the talent and skills her peers bring to the gym, Emily does not find a warm welcome; Kaylie and particularly Lauren take issue with Emily's arrival. While Payson, Kaylie, and Lauren had always been the three top gymnasts at The Rock, Emily's arrival alters this balance and challenges Lauren's position on the podium. While the additional competition will eventually lead to improvements all around, the process is showcased as a class struggle.

Conventions of the genre dictate that the Cinderella narrative would make Emily a winner, yet *MioBi* resists such a narrative and removes Emily from competition and from the show before the end of season two, to have a family.[24] The topic of teen pregnancy brings to the fore the question of control over one's body and life, and further entrenches cultural characteristics of different classes: Emily's pregnancy limits the scope of the Kmetkos' future as it echoes her mother's (Chloe, played by Susan Ward) own decisions. Chloe stands out and further affirms this outsider status, as she is "made-up" and wears cheap, tight dresses and high heels, when the other mothers at the gym tend to wear jeans and sneakers. She too is kept to the margins of The Rock's social group, but contrary to Emily, Chloe appears totally comfortable with who she is. If the inter-class fighting on the mat and on the sidelines were not sufficient, the class struggle comes to a head when Lauren's father and Emily's mother become romantically involved. Needless to say, they are, on the surface, not a good match, but the relative success of their relationship (they date for most of the second half of season one) suggests that *MioBi* writers are interested in exploring the complexity of social divisions and their absurdity. Removing Emily from the competition for teen pregnancy entrenches her within her socioeconomic background, but also removes the motive for the competition between the lead athletes at The Rock. To sustain the competitive edge driving the narrative, a new outsider (Jordan Randall, played by Chelsea Tavares) is brought in for the final eight episodes.

Further developing the layers of class distinctions, *MioBi* focuses on the challenges brought on by medical expenses on middle class families such as the Keelers. Payson's family offers further insights into the complexities of privilege. Following a serious injury (discussed below), her parents (Kim and Mark, played by Peri Gilpin and Brett Cullen) struggle to make ends meet with the onslaught of Payson's medical bills that insurance will not cover. While they try and hide their financial situation from Payson and her sister (Becca, played by Mia Rose Frampton) at first, her father's move back to Minnesota for work, as well as their serious consideration of selling their home, soon make it abundantly clear to Payson that the sacrifices her family has made for her career are more in line with Chloe's sacrifices for Emily than either Lauren or Kaylie's experiences.[25] Payson, like Emily, is impelled to try and help her family out by offering to go professional at varying points throughout the series; her parents finally succumb to Payson's requests to sign with a sponsor, but as a result, she loses her chance at a NCAA college gymnastics scholarship.[26] The closeness of the lower and middle classes are highlighted especially well in a scene in season one, when all the mothers come together to support Payson's mother, but only one of them truly understands the struggle the Keelers are facing: Chloe Kmetko, who shares "the perilous specter of immobility," both literally and socioeconomically.[27]

Societal Issues: Body Image

In her essay, Lindner makes a distinction between a believable and fictional athleticism. Bodies like those of Hayley's in *Stick It* (and, we argue, more so yet, all of the *MioBi* girls) are believable—though the muscles they exhibit do not fit in girly representations common in neo-feminist media: "the gymnasts' bodily exertion and the bodily effort and strength that are required in order to make the gymnastics performances *appear* effortlessness, are overtly emphasized in the film's training sequences."[28] Neo-feminist media favors disciplined bodies, but bodies that will affirm gendered boundaries and not confuse them: excess muscle will therefore be erased so as not to confuse male and female genders.[29] *MioBi* invests in the staging of the efforts required as well as the time it takes for these muscles to build. Within the constraints of 103 minutes, *Stick It* gives a relatively realistic framework for its character's physical development of strengths; paying lip-service to the toning, however, serves more as a narrative device (Haley's punishment and makeover), than fully as a means for the audience to conceive of the degree of efforts required.

Furthermore, much is made in *Stick It* of physical appearance: between the aesthetic choice of showing Haley plunge and exit an ice bath partially naked twice, and that of dressing her not in a leotard, but in shorts and a sports bra so as to show off her midriff, *Stick It* objectifies the gymnasts in ways that *MioBi* resists. This objectification is also the focus of jokes in Bendinger's film: in one instance, fellow gymnast Joanne (Vanessa Lengies) repeatedly asks for the team to choose a new leotard, preferably one that would allow for her arm muscles to show, because "she has a right to bare arms." Girls in *MioBi* are also shown in intense hours of muscle toning exercises, and regularly demonstrate their physical prowess, but they are in realistic attire—leotards, tee-shirts and/or sweatpants—thereby highlighting the underdevelopment and prepubescent bodies of gymnasts, not constructing them as sexually available.

Nevertheless, because gymnastics, one of the sports that Lindner refers to as "female-appropriate" (including figure skating and competitive dance), is one of the sports that sees participants score points by impressing a panel of judges, there is a built in element of spectacle.[30] Bodies are constantly under scrutiny in *MioBi*; girls spend their days in the gym as one would at a day job and do so six days a week building and toning muscle. They are to watch what they eat, which is emphasized by outfitting The Rock with a scale and height rod to monitor the athletes' growth, as well as dialogue ("that's full of carbs") and the narrative itself, which hinges on issues related to body image.[31] The characters of Kaylie and Payson are particularly interesting to examine regarding the issue of body image, as both of these characters are vehicles for unique messages. On the one hand, Kaylie's character is used as an anti-eating disorder vehicle, and Payson's, on the other hand, serves to facilitate ease with a changing body, both natural for a prepubescent girl and unnatural due to injury, and how this may affect one's aspirations and dreams.

Although Kaylie seemingly never wants for anything, the pressure she and her parents—sometimes consciously, sometimes not—place upon her to be number one is mentally and physically crushing and leads her to take control of her body in self-destructive ways. Kaylie might arguably be under the most strain, because of her parents' personal success in highly public careers of their own. Her mother (Roni, played by Rosa Blasi) had a brief professional singing career, and her father (Alex, played by Jason Manuel Olazabal) is a retired Major League Baseball star. Their accomplishments visibly and verbally surround Kaylie on a daily basis. Her mother still keeps the recording studio in their home, and her father has baseball memorabilia strewn throughout the house and refers to it constantly.[32] From the very first episode, references are made to gymnasts having to be conscious of their food and diet choices. However, the more Kaylie grows in her competitive career, the more her parents' relationship falls apart, and the more she becomes obsessed with her weight. Eating little and exercising more leads her to feel faint, because her body cannot keep up any longer.[33]

More than a simple turn in the narrative of *MioBi*, Kaylie's anorexia is highlighted within the narrative and outside of it; it is the focus of inserted short commercials, akin to product placement. Indeed, at choice moments in the television show, when the suspense is at its utmost, the narrative stops with a cut to Kaylie, standing in her parents' study, facing the camera and speaking into it, breaking the fourth wall, telling the audience that "having an eating disorder is a complicated issue, and nothing to be ashamed of. But the first step to recovery is talking about it. To share your story, or for help, contact the National Eating Disorders Association at 1-800-931-2237. The life you save could be your own." While she says these lines, a box appears at the bottom of the screen with the same telephone number and the fully spelled out name for the national organization fighting eating disorders.[34] This, in structure and composition, is a short commercial break, but built in the television show itself and fully integrated in it, insomuch as it makes use of the set and of one of the characters. The message itself also echoes *MioBi*'s narrative: Kaylie takes control of the disorder when she rips out a page in her journal with calorie count notes and replaces it with "My name is Kaylie Cruz & I am anorexic."[35] While she has yet to share her story, as she calls viewers who might be suffering from the same disorder to do in the commercial, she is putting it in words and telling herself, thereby acknowledging that indeed she is suffering from a disorder that she must overcome. Kaylie comes to this life-saving realization only after she loses her friend Maeve (Alice Greczyn) to the disorder. This turn in the narrative also marks a new beginning for her gymnastics. More intentional in her choices, Kaylie, from this point on, controls her career like she does her body, by fully deploying her newfound agency. The emphasis on the struggle, as well as the self-inflicted starvation and eventually the hinted-at resolution of the problem by its acknowledgment, and more efforts, all resist classic neo-feminist media, and suggests a new order.

While success for Kaylie is a very public event (after winning Nationals, Boulder declares "Kaylie Cruz Day"), it also creates the pressure we described above.[36] In that single episode, the power dynamic changes. Her relationship with Payson especially suffers, because Payson was supposed to be the golden girl and win it all. Instead, Payson fractures her back falling from the uneven bars at Nationals, and doctors tell her she will never do gymnastics again. While her back injury might be the most egregious physical hurdle Payson encounters in the show, it is not the only one. From the very first episode, Payson is presented as a powerhouse: she is the best gymnast, yes, but that is so because of her single mindedness and the strength of her execution. What she is shown to lack, however, is grace. While she promptly dismisses such criticism as unimportant, the realization that she is losing points because of it, forces her to reconsider. Payson is led to come to terms with her femininity in the fourth episode of the series. It is much more than simply getting her to be more graceful (she argues that her strength is her power, not her grace; the girly stuff is Kaylie's thing), however, it is also getting her, and her mother, in touch with the ways in which they are feminine. The plot device to achieve this is a mother/daughter fashion show. The relationship amongst girls, and between them and their mothers is showcased here in order to highlight the construction of various femininities (as defined above), but also the ways in which young girls are to develop and appropriate their own. This very didactic episode centers on Payson as she comes to terms with the fact that femininity is not necessarily contradicting strength, and in fact can be whatever she decides to make it. *MioBi* in this instance makes good use of the visual tradition of neo-feminist media by utilizing clothing as a metaphor for diverse femininities. However, the clothing is not owned by the gymnasts, let alone purchased by them; rather, the clothes are being auctioned as a fundraiser.[37]

While this episode might have been the first time the question of femininity is addressed up front in *MioBi*, it is overtly picked up in season two, when Payson is back on the mat after her surgery and, against all predictions, is actually competitive again.[38] However, because her body has changed, her strength may not be power any longer so much as it is grace and the artistic quality of her performance. While she never thought of herself as a graceful gymnast, Payson comes to embrace such qualities as compelling thanks to a ballet that Sasha Belov (Neil Jackson), The Rock's coach, invites her to see. Significantly, he takes her to experience a reinvented version of Tchaikovsky's *Swan Lake*, a classical ballet centered on the story of a woman who is reborn as a swan. As Sasha explains to Payson during the performance, this ballet is about coming to terms with change. And indeed, Payson does come to terms with the changes of her body, and the needed adjustments to her gymnastics. She is enthralled not only by the dancer's grace, but also by her power, and understands that there is more to elegance, artistry and grace than meets the eye. Convinced that this is not a vacuous endeavor, but instead a meaningful one, she subsequently returns to the gym wholly invested in developing her full extensions, and succeeds in conveying a complete story via her body, a feat she never thought she would muster in the first season of the show. It is this drastic change that earns her a place on the National Team by the middle of season two.[39] Payson succeeds, and while this is something of a Cinderella moment in *MioBi*, it is also a lesson in adaptability. While some women are blessed with it all (Kaylie is repeatedly referred to in this manner), others not only have to work at their success, but sometimes also have to reinvent themselves and adapt to the changes life imposes. Payson does so, not without efforts, but with a devotion that sets her apart from most, including her fellow Olympic athletes in the show.

Conclusion

ABC Family's production *Make It or Break It* contributed an empowering visual message to young, and older, viewers of the show over the three years of its airing on the network, and it continues to be available through streaming and rental services like Hulu and Freeform's own website. The show also experienced a bit of media resurgence related to coverage of women's gymnastics at the 2016 Rio Olympics.[40] As the series ends with the girls all achieving their dream of making it onto the USA Olympic team,[41] the message is clear: hard work and sacrifice are necessary but, perhaps, even more important is the support of other team members that helps you overcome every obstacle to attain your goal.

In this essay, we have shown that ABC Family contributed in *MioBi* a television show that challenges more common teen girl media, most of which feed into postfeminist culture. While few feminist media scholars will deny the compelling aspect of much of the postfeminist cultural productions, shows like *MioBi* stand out as not only teetering on the brink of postfeminist media's appeal, but more so affirming some of the limitations of the postfeminist media. By resisting major aesthetic, narrative, and generally political stances of postfeminist media, *MioBi* offers empowerment, where others might offer "girl power." It does not just repeat slogans of the sides of feminist messages it wants to showcase, but instead discloses aspects of the power struggle that feminist activists and scholars are still working to overcome. The narrative focus on a class-based depiction of society is a key element in this. *MioBi* does not gloss over socioeconomic differences as we showed earlier, but shows what need is, what circumstances might bring about these needs, and overall, resists the romanticization of poverty.

Another aspect *MioBi* does not romanticize is the pain involved in preparing one's body to execute gymnastics stunts. Far from the difficulty some might have throughout the filmography and televisiography of post-feminist media in choosing one's outfits, or identifying the right partner for another high school social, the four lead protagonists of *MioBi* suffer from anorexia, or, on the other hand, the challenges one faces when recovering from a major bodily injury. From the very first episode on, *MioBi* puts bodies front and center, but in ways that come to challenge Hollywood's objectification of teen girls' bodies. While most elite gymnasts will probably echo Payson's statement "our bodies are everything we have. Everything we work for,"[42] she is knowingly commenting on a professional investment in the gymnasts' bodies that goes well beyond looks, social pressure, and making oneself the object of male desire. Accordingly, *MioBi* invests in realism where neo-feminist media would instead invest in refined aesthetics. This focus on realism starts with the countless hours of training to sculpt the gymnasts' bodies as the main focus of the show (instead of concealing the efforts), more natural lights, and a camera that frequently sets the stage (in the gym, in the gymnasts' homes, or in one of the two restaurants) and locates the teen girls in their environment, however rough that might be. Part of this realness is achieved by way of actual former competitive gymnasts performing the stunts, including regulars such as Natalie Padilla, Tarah Paige, Ariana Berlin, and Renae Moneymaker.[43] The realism of *MioBi* is not an artistic attempt, but instead another means by which it stands out in the landscape of neo-feminist media productions.

Breaking again with the focus on fantasy of postfeminist media is the distance *MioBi* takes from the Cinderella narrative. Emily is a perfect Cinderella in this show: the rags to riches narrative was likely expected by all members of the audience, and rightly so. The first two seasons of *MioBi* play with our expectations: will Emily make it? Will she finally be recognized as the princess she ought to be? But *MioBi* resists simpler narratives to develop a complex character struggling to belong and overcome the class struggle she is mired in despite her best efforts. *MioBi* develops Emily's character as one whose wanting childhood leads her to break rules so as to be able to provide for her family: she works despite The Rock's prohibition against employment, and she finds it difficult to trust anyone since she has never been able to rely on anyone other than herself. It is her trust in someone other than herself that eventually leads her to commit to gymnastics, her peers, and Sasha, in ways she could not have otherwise. The trust she placed in Sasha led her to win his, in turn, and leads her to a spot on the USA National Team. Significantly, this was an unexpected feat for a competitor in her position, both from a socioeconomic standpoint, and as a newcomer to professional training. While this exemplifies Lindner's criticism of the limitations of the exclusive relationship women establish with their coach in sports films, it also further emphasizes the collective aspect of successes such as Olympic wins. Contrary to neoliberal dogma highlighting individuals' achievements, *MioBi* puts great emphasis on the team effort and the support girls provide one another. If Lauren is the only one who never truly will be on team Emily, all the other protagonists understand that their bond is their sole source of power. Neither their individual talents, strength, or grace will allow them to break through the complex world of gymnastics at the national and international level.

NOTES

1. Nellie Andreeva, "ABC Family's 'Make It or Break It' to End," *Deadline Hollywood*, April 26, 2012, http://deadline.com/2012/04/abc-familys-make-it-or-break-it-to-end-262399/.

2. Emily Kmetko is replaced by Jordan Randall (Chelsea Tavares) in season three.

3. According to Hilary Radner, two distinct characteristics define "girly." On one hand, it refers to media made for a feminine audience that showcases female characters in the process of changing and improving. In her words, "a girl is always in the process of 'becoming'" (7). The second aspect refers to the move for women to reclaim a term that referred to magazines and other media geared toward masculine audiences in the 1940s and 1950s representing sexually available women. This move is thus empowering. While Radner borrows the term from Charlotte Brunsdon, who coined it, she distinguishes her take insomuch that for her, girly media are neo-feminist in nature; for Brunsdon, they are postfeminist. In this essay, we will not engage in the distinction between neo- and postfeminism. However, we take issue with Radner's claim that all girly media is always already neo-feminist on the grounds that *Make It or Break It* is a good example of media showcasing girls in the process of becoming women without engaging in the consumerist paradigm of neo-feminism. See Hilary Radner, *Neo-Feminist Cinema* (New York: Routledge, 2011). Brunsdon coined "girly" in her essay "Post-Feminism and Shopping Films," in *Screen Tastes: Soap Opera to Satellite Dishes* (New York: Routledge, 1997), 81–102.

4. *Stick It*, directed by Jessica Bendinger, Burbank, CA: Disney (Buena Vista), 2006.

5. "Sunday, Bloody Sasha, Sunday," *Make It or Break It*, season 1, episode 4, ABC Family, July 13, 2009.

6. Wyatt Smith plays the part of Emily's brother Brian. Although his character uses a wheelchair, Smith is, in fact, able-bodied, a controversial casting decision. For representative criticism of able-bodied actors playing the part of physically disabled characters, see David Kociemba, "'This Isn't Something I Can Fake': Reactions to *Glee's* Representations of Disability," *Transformative Works and Cultures* 5 (2010), http://journal.transformativeworks.org/index.php/twc/article/view/225/185.

7. Payson deals with her own disability for much of the second half of season 1.

8. Though our essay focuses primarily on gender, we acknowledge that *MioBi* centers whiteness in obvious and uncomfortable ways. *MioBi* is a television show that focuses on a mostly racially diverse sport (gymnastics). For example, the USA Gymnastics women's team for the 2016 Olympics in Rio included racial minorities in three of the five spots: Gabby

Douglas (African American), Simone Biles (African American), and Laurie Hernandez (Latina). In *MioBi*, the exclusion of racial diversity from its cast reinforces a system of domination, to affirm racial, sexual, and ability (among others) norms and warrants further analysis by critical media scholars. Classic contributions to the field of whiteness studies include Richard Dyer, *White* (New York: Routledge, 1997); Ruth Frankenberg, *Displacing Whiteness* (Durham: Duke University Press, 1996); and George Lipsitz, *The Possessive Investment in Whiteness* (Philadelphia: Temple University Press, 1998). More recent studies include Rake Shome, "Outing Whiteness," *Critical Studies in Media Communication* 17, no. 3 (2000): 366–371; or Joy Taylor, "'You Can See Me,' or Can You?: Unpacking John Cena's Performance of Whiteness in World Wrestling Entertainment," *Journal of Popular Culture* 47, no. 2 (2014): 307–326.

9. Yvonne Tasker and Diane Negra, eds., *Interrogating Postfeminism: Gender and Politics of Popular Culture* (Durham: Duke University Press, 2007), 19.

10. Tasker and Negra, *Interrogating Postfeminism*, 4.

11. Radner, *Neo-Feminist Cinema*, 6.

12. Helen Gurley Brown, *Sex and the Single Girl* (New York: Bernard Geis Associates, 1962); Betty Friedan, *The Feminine Mystique* (New York: Dell, 1963).

13. Radner, *Neo-Feminist Cinema*, 17–22.

14. Celestino Deleyto, *The Secret Life of Romantic Comedy* (Manchester: Manchester University Press, 2009), 153.

15. Two exceptions to this include *Bring It On* (2000) and *Bend It Like Beckham* (2002), two films compared for their takes on female competition in cheerleading and soccer respectively, in Roz Kaveney, "On Being Good at Things: Female Competence and Sexuality," in *Teen Dreams: Reading Teen Film from Heathers to Veronica Mars* (New York: I.B. Tauris, 2006), 161–176. We also must acknowledge the growing scholarship in feminist sports sociology, a vibrant field in the UK. Much of that scholarship focuses on women's position in sports considered the purview of men (soccer, rugby), and in more recent years, on their racial and ethnic identities. Key contributors to this field include Jayne Caudwell and Sheila Scraton.

16. Katharina Lindner, "Blood, Sweat and Tears: Women, Sport and Hollywood," in *Postfeminism and Contemporary Hollywood Cinema*, ed. Joel Gwynne and Nadine Muller (London: Palgrave Macmillan, 2013), 242–243.

17. *Ibid.*, 239.

18. *Ibid.*, 242.

19. Ann Chisholm, "Acrobats, Contortionists, and Cute Children: The Promise and Perversity of U.S. Women's Gymnastics," *Signs* 27, no. 2 (2002): 434–440.

20. "Sunday, Bloody Sasha, Sunday," *Make It or Break It*.

21. Tasker and Negra, *Interrogating Postfeminism*, 2.

22. Emily's socioeconomic status frames every aspect of her narrative and is brought to the foreground on numerous occasions throughout the series. Jennifer Hargreaves argues that class has been proven to have an enormous impact on teen girls' expectations for their sporting futures, so it is not surprising that Emily consciously questions her motivations for devoting so much time to gymnastics. Middle-class girls (and one may presume upper-class, too) visualize their adult selves as being educated with access to leisure time, and so invest more time and energy in adolescent sports. By contrast, "working-class girls tend to visualize their futures as being limited by marriage, housekeeping and child-rearing, and their attitudes to leisure and sports are influenced by these ideas" (Jennifer Hargreaves, *Sporting Females: Critical Issues in the History and Sociology of Women's Sports* [New York: Routledge, 1994], 156–157). And this is precisely where Emily ends her gymnastics career, choosing child-rearing instead when she becomes pregnant, a decision that the other girls (from middle and upper-class families) have difficulty understanding. "Requiem for a Dream," *Make It or Break It*, season 2, episode 16, ABC Family, May 2, 2011.

23. Further examples of this class dynamic include the following exchanges: Lauren says to Emily, "You don't have to play little girl from the other side of the track." Conversely, Emily tells Lauren when their parents' relationship becomes more public, "It'll be a cold day in hell before my family takes as much as a postage stamp from your father." "Friends Close,

Enemies Closer," *Make It or Break It*, season 2, episode 1, ABC Family, 28 June 2010. When Emily finds out Lauren's dad is the one behind her so-called "Kipman Group" sponsoring her scholarship, she works doubly hard to pay him back. She does not want to owe the Tanner family anything. "What Are You Made Of?" *Make It or Break It*, season 2, episode 7, ABC Family, August 10, 2010; "Rock Bottom," *Make It or Break It*, season 2, episode 8, ABC Family, August 17, 2010.

24. "Requiem for a Dream," *Make It or Break It*, season 2, episode 16, ABC Family, May 2, 2011.

25. "The Buddy System," *Make It or Break It*, season 2, episode 13, ABC Family, April 11, 2011.

26. "What Lies Beneath," *Make It or Break It*, season 2, episode 19, ABC Family, May 23, 2011; "Worlds Apart," *Make It or Break It*, season 2, episode 20, ABC Family, May 23, 2011.

27. Tasker and Negra, *Interrogating Postfeminism*, 19. The episode involves the mothers ensuring that Kim Keeler wins the $10,000 Bingo prize to save the Keeler family home. See "The Great Wall," *Make It or Break It*, season 1, episode 18, ABC Family, February 22, 2010.

28. Lindner, "Blood, Sweat and Tears," 251. Chisholm also addresses the importance of illusion of ease when gymnasts perform in competition, especially from the perspective of televised productions. See Chisholm, "Acrobats, Contortionists, and Cute Children," 418–419.

29. Tasker and Negra, *Interrogating Postfeminism*, 19.

30. Lindner, "Blood, Sweat and Tears," 245.

31. "Pilot," *Make It or Break It*, season 1, episode 1, ABC Family, June 22, 2009.

32. People who meet Alex Cruz and/or are trying to impress him also reference his baseball success. For example, when Austin has dinner with Kaylie and her parents, he brings up all the baseball highlight clips he has been watching of Mr. Cruz on YouTube. "Growing Pains," *Make It or Break It*, season 3, episode 4, ABC Family, April 16, 2012.

33. "Friends Close, Enemies Closer," *Make It or Break It*, season 2, episode 1, ABC Family, June 28, 2010.

34. "All or Nothing," *Make It or Break It*, season 2, episode 2, ABC Family, July 5, 2010.

35. "Life or Death," *Make It or Break It*, season 2, episode 14, ABC Family, April 18, 2011.

36. "The Eleventh Hour," *Make It or Break It*, season 1, episode 11, ABC Family, January 4, 2010.

37. "Like Mother, Like Daughter, Like Supermodel," *Make It or Break It*, season 1, episode 5, ABC Family, July 20, 2009.

38. "And the Rocky Goes To...," *Make It or Break It*, season 2, episode 4, ABC Family, July 20, 2010.

39. "At the Edge of the Worlds," *Make It or Break It*, season 2, episode 10, ABC Family, August 31, 2010.

40. Megan Garber, "*Make It or Break It* Is Pre-Olympics Gold," *The Atlantic*, July 11, 2016, http://www.theatlantic.com/entertainment/archive/2016/07/make-it-or-break-it/490 757/. Some cast members also reunited on August 10, 2016, the day after the U.S. "Final Five" women's gymnasts won team gold in Rio, for a Facebook live chat to reconnect with fans. Deepa Lakshmin, "The Make It or Break It Cast Reunited for a First Place Photo," *MTV News*, August 11, 2016, http://www.mtv.com/news/2918218/make-it-or-break-it-reunion/.

41. "United Stakes," *Make It or Break It*, season 3, episode 8, ABC Family, May 14, 2012.

42. "Sunday, Bloody Sasha, Sunday," *Make It or Break It*.

43. See IMDb, "Series Stunts," for a complete list of stunt doubles. "Make It or Break It," *Internet Movie Database*, January 29, 2017, http://www.imdb.com/title/tt1332030/?ref_= tt_ov_inf.

"We are definitely not the Brady Bunch"

An Analysis of Queer Parenting in the Teen Family Drama The Fosters

Stephanie L. Young *and* Nikki Jo McCrady

"So, you're dykes," says a caustic Callie, the new foster daughter. "They prefer the term 'people.' But yeah, they're gay," replies Jesus, Stef and Lena's adopted son. "And he's the real son," Callie points to Brandon, Stef's biological son. Stef nervously laughs as the rest of the family awkwardly sit in silence at the kitchen table.[1] With Callie's defensive remarks, audiences are not only introduced to the "unconventional family" of *The Fosters*, but also encouraged to reflect on their own definition of and biases regarding what makes a family. Indeed, as Montgomery Jones observes, "Callie's ignorance of LGTBQ people and her own experiences with the foster care system put her outside of the family dynamic immediately—she can't even really see that there *is* one."[2]

Premiering in 2013 on the Freeform network (originally ABC Family), *The Fosters* is a family drama that focuses on an interracial lesbian couple, Stef Foster (Teri Polo) and Lena Adams (Sherri Saum), and their five children. According to Mike Hale of *The New York Times*, *The Fosters* is "a multiple threat, wrapping up gay parenting, blended families, adoption and the foster care and juvenile justice systems."[3] Media critics have praised the television series for its depiction of gay parenting and complex queer characters.[4] As "prime-time television's most realistic same-sex union,"[5] the series depicts Lena and Stef as multidimensional lesbian characters who resist stereotypes of "the hyper-sexualized 'lipstick lesbian' or that of the man-hating and entirely un-feminine 'butch.'"[6] As Sarah Caldwell of *Entertainment Weekly* states, "Seeing a lesbian, biracial couple on a family TV show is a big deal."[7]

Like many of Freeform's programs, *The Fosters* is aimed at teenage audiences and addresses challenging issues such as teen pregnancy, sexual assault, sexuality, and the foster care system. What makes *The Fosters* unique in the landscape of family programming is that it not only reflects "meaningful diversity" with its fully realized characters that portray race and sexuality in developed ways,[8] but it also provides audiences with a diversity of kinship ties. The series not only depicts a queer interracial family, in the sense that it is headed by a lesbian couple, but also incorporates a variety of parent-child relationships: Brandon (David Lambert), who is Stef's biological son from a previous marriage with Mike Foster (Danny Nucci); the adopted twins Mariana (Cierra Ramirez) and Jesus (Jake T. Austin); and Callie (Maia Mitchell) and her half-brother Jude (Hayden Byerly), who are fostered and then adopted. As *The Fosters'* executive producer Jennifer Lopez explained, she wanted the show to reflect "a diverse society. While our family compositions might differ, the experiences of a family are universal."[9] As such, *The Fosters* depicts a corporate branding of "cultural diversity" that, along with racial diversity, incorporates the "inclusion of LGBTQ characters in its vision of 'a new kind of family' is the most explicit challenge to traditional family values."[10]

Additionally, television has seen an increased representation of LGBT characters, particularly same-sex partners and queer families, with shows like HBO's *Six Feet Under* (2001–2005), ABC's *Modern Family* (2009–present), FOX's *Glee* (2009–2015), ABC's *Brothers and Sisters* (2006–2011), and NBC's *The New Normal* (2012–2013).[11] Interestingly, however, gay parenting has been often centered on gay fatherhood rather than lesbian motherhood. As *The Fosters'* co-creator Bradley Bredeweg noted, "We started looking around at the landscape and thought about maybe telling a story about the American family with gay dads, but we felt that had been done a few times before and rather well. Then we realized that there was a kind of vacuum when it came to stories about women raising families."[12] As Renee Fabian observes, the representation of lesbian mothers is groundbreaking, and "for LGBT youth drawn to the show by *The Fosters'* kids and their stories, they also have the benefit of seeing Stef and Lena in a stable, loving relationship."[13]

As gay parenting becomes increasingly visible in contemporary American media, it is important to explore how these portrayals of motherhood (and fatherhood) are enacted. This essay, then, critically examines the representation of lesbian motherhood and the construction of a queer family on the series *The Fosters*. Drawing upon queer and gender studies, we explore how *The Fosters* provides an alternative vision of family while reinforcing traditional family structures and heteronormativity in America. Specifically, we look at how Stef and Lena's relationship plays a crucial role in the redefinition of family.

Television and the Representation of LGBT Characters

Much research has examined the representation of gay, lesbian, bisexual, and transgender characters on television and film.[14] While there has been increased visibility of queer identities, television has perpetuated and continues to perpetuate a number of stereotypes, including the sexually promiscuous bisexual[15] and the flamboyant gay man as comedic relief.[16] Additionally, queer characters often have been portrayed as isolated, maladjusted, and disconnected from their families. Vito Russo's formative work, *The Celluloid Closet: Homosexuality in the Movies*, traces the history of American film's depictions of gays and lesbians, finding them often cast as "pathological, predatory, and dangerous."[17] As Frederick Dhaenens notes, queer characters are often located within two tropes—the villain or the victim.[18] Indeed, scholars have noted that "gay lives are narrated through 'the harder path' and 'the miserable life,'"[19] and "mainstream images of gayness perpetuate a tragic cycle of punishment and exclusion."[20]

Lesbians, too, have been stereotypically feminized in popular media, reinforcing the "lipstick lesbian" or "femme" fantasy of heterosexual audiences.[21] The television lesbian is often constructed as the "gentle, sensitive, soft-hearted, soft-spoken, absolutely non-butch" woman.[22] However, this "user-friendly lesbian"[23] is problematic as it erases the diversity of lesbian identities and heterosexualizes lesbian bodies.[24] Even when shows do depict more masculine lesbians, with Shane from Showtime's *The L Word* (2004–2009) or Leda from Showtime's *Queer as Folk* (2000–2005), butch characters are viewed as hypersexual and a threat to gender norms.[25]

With the increased visibility of lesbian characters on television shows, scholars have been critical of these depictions. For example, Katerina Symes explores how the character of Piper Chapman on Netflix's *Orange Is the New Black* (2013–present) is used as a "heterosexual proxy" for straight-identified female audiences to explore queer sexualities.[26] Pei-Wen Lee and Michaela D. E. Meyer problematize the depiction of lesbian identities on *The L Word*, noting that absence of non-queer relationships and interactions with heterosexual individuals constructs an insular world that "perpetuates an ideology of avoidance."[27]

Interestingly, teen programming has been at the forefront for introducing audiences to complex gay, lesbian, bisexual, and transgender characters with shows such as The WB's *Dawson's Creek* (1998–2003)[28] and *One Tree Hill* (2003–2012).[29] Much of the prime-time television series *Glee,* which merges the genres of the musical comedy and teen drama, provides audiences with a diversity of queer teen sexualities.[30] In fact, studies have looked at the effects portrayals of queer characters have on audiences, noting that positive representations can shape the knowledge and attitudes of young people, particularly on discussions of sexuality.[31]

In addition to incorporating LGBT characters in teen dramas and the depiction of "chosen" families, queer communities, and non-normative kinship ties,[32] one growing trend on network and primetime television has been the representation of gay parenting.[33] Queer research has analyzed same-sex parenting, particularly within the ABC series *Modern Family.* Peter Kunze explores how *Modern Family* not only normalizes same-sex parenting, but also offers a critique of heterosexual masculinities.[34] Steven Edward Doran also examines how *Modern Family,* like many television shows, utilizes the strategy of "homodomesticity," or featuring gay and lesbian characters in traditional domestic roles and contexts to assimilate queer identities in the mainstream. As he explains, "the performance of homodomesticity requires erasing, overlooking or suppressing those elements of gay identity that could potentially disrupt the smooth operation of heteronormativity."[35] In fact, homodomesticity reflects what Michael Warner identifies as the "politics of normal" in which one is "to blend in, to have no visible difference and no conflict."[36]

Andre Cavalcante notes that while television engages in the normalization of queer parenting narratives, these representations are limiting as they often present the "ideal type" of gay parents as white, upper-middle class, gay men, thus privileging white masculinity to "manage the anxiety of social difference."[37] Additionally, Suzanna Danuta Walters identifies three primary tropes related to contemporary depictions of gay and lesbian families: (1) the "universalist" theme that normalizes gay families, (2) the "still abject adolescent" theme that addresses homophobia but "assures us that progress is most assuredly being made," and (3) the "alternative family" theme that commends LGBT "brave new families."[38]

In many ways, television's depiction of gay couples and gay-headed families reflects the larger acknowledgment of the diversity of family types in America—"families headed by a divorced parent, couples raising children out of wedlock, two-earner families, same-sex couples, families with no spouse in the labor force, blended families, and empty nest families."[39] However, while there are shifting familial trends, with an estimated six million Americans who have LGB parents,[40] gay parenting still challenges the conceptualization of the "traditional" nuclear family as fundamentally heterosexually monogamous and founded on blood ties. As such, "lesbian mothers and gay fathers challenge the normative discourse of genetic parenthood (that everybody has a right to know their 'genetic origins')."[41] Gay and lesbian couples often form families with children in multiple ways, including utilizing reproductive technologies such as in vitro fertilization and surrogacy, engaging in co-parent or joint adoption, and fostering children. Indeed, films such as *The Kids are All Right* (2010) center on narratives of donor conception in relation to queerness and normative reproduction.[42]

There also has been increasing research in family studies with regard to lesbian motherhood and parenting.[43] Scholarship has observed how lesbian mothers may utilize dominant discourses of maternity to frame their identities[44] and the unique challenges lesbian mothers face, including the stereotype that children may suffer negative emotional consequences due to the lack of a father.[45] However, these normalizing discourses erase significant differences between same-sex and heterosexual couples in terms of division of labor, conflict management, and communication,[46] reinforcing couplehood, marriage, and family in heteronormative ways.[47]

In sum, much research has examined LGBT characters on television as well as the unique challenges gay and lesbian encounter in parenting. However, little media research has focused on the representation of lesbian-headed households. As television continues to create space for gay sexuality and romantic queer relationships, *The Fosters* as a teen family drama can provide us with a case study for exploring how lesbian motherhood and family are represented.

Methodology

The Fosters provides audiences with a complex representation of a "nontraditional" lesbian-headed family. By engaging in a close textual analysis of the series, we examine the ways in which the characters, dialogue, and storylines both challenge and reinforce dominant heterosexual discourses about family. Specifically, a close reading allows the television critic to analyze visual and verbal components communicated by the text. Close textual analysis requires the critic to "linger over words, verbal images, elements of style, sentences, argument patterns, and entire paragraphs and larger discursive units within the text to explore their significance on multiple levels."[48] As Jennifer Dunn and Stephanie Young explain, "the purpose of textual analysis is to systematically describe and interpret the characteristics of the content, structure, and/or form of symbolic communication."[49] Thus, textual analysis is a valuable method for revealing layered cultural meanings and ideologies within a text through the systematic breakdown of various elements and examination of how these elements function together.[50]

In addition, we draw upon feminist media criticism to provide a critical lens for examining these discourses of family. As Dunn and Young further state, "a feminist perspective can provide a critical lens for examining how popular culture communicates gender norms, roles, behaviors, and stereotypes for women and men."[51] By taking up a feminist perspective, we look at how constructions of gender and sexuality, including LGBT issues, are intertwined within the context of familial relationships. Brian Ott and Robert Mack find that "feminist media scholars understand media texts as products of sexist social systems, and they look especially at the ways in which patriarchal systems of power inform the creation of media texts."[52] By taking a feminist approach, we examine how lesbian parenting is depicted in *The Fosters*—how messages about sexuality, gender, and family are communicatively constructed through and with the relationship between Stef and Lena. In fact, interpersonal relationships between characters on television can be viewed as a central unit of analysis for textual criticism.[53]

Specifically, we each viewed and analyzed the first 42 episodes of the series. We focused on the first two seasons, which originally aired from June 3, 2013, to March 23, 2015, as they familiarized audiences with the relationship between Stef and Lena and were aired prior to the Supreme Court ruling to legalize same-sex marriage.[54] While we recognize that there is not a causal link between the airing of these episodes and marriage equality, we believe it important to see how the series crafted a message that normalized lesbian couplehood prior to federal legalization of same-sex marriage.

While viewing each episode, we took extensive notes on messages related to parenting and relationships within the family. Then we compared notes and identified patterns associated with characters, dialogue, and storylines related to a lesbian-headed family. Ultimately, four major themes emerged: (1) negotiating tradition within contemporary norms, (2) normalizing sexual intimacy, (3) managing parental roles, and (4) reframing discourses of family.

Negotiating Tradition within Contemporary Norms

Throughout the series, the lesbian-headed Fosters is normalized as a "real family" by depicting them engaging in various familial rituals—Mariana's quinceañera, Stef and Lena's wedding, the funeral of Stef's father Frank Cooper (Sam McMurray), and Jude's adoption "birthday" celebration. Family rituals, be they culturally prescribed practices or more unique traditions of the family, are "a symbolic form of communication" that play a significant role in creating and maintaining a sense of familial identity and belonging.[55]

However, ritual is rooted in tradition (often religious) or "cultural beliefs and practices" that "seem ordained in the order of things"; that is, they are the taken-for-granted, "natural" ways of life (e.g., one should get married).[56] In many ways, *The Fosters* attempts to simultaneously reinforce and re-envision tradition in contemporary ways. For example, in "Quinceañera"[57] Stef and Lena decide to throw Mariana[58] a quinceañera, a traditional Latin American celebration of a girl's fifteenth birthday and a rite of passage into womanhood.[59]

The episode begins with Mariana being fitted for her quinceañera dress, accompanied by Lena and Mariana's friend Lexi (Bianca A. Santos). When Stef arrives, she quickly kisses Lena on the cheek. The dressmaker, a Latina woman, is wide-eyed and awkwardly smiles. Then, the scene cuts to Mariana and Lexi giggling as she undresses. "Did you see her face when Stef walked up in her uniform? Hilarious!" says Lexi. "I hate when people say *bueno*, like you need their approval for something. When you tell people you have a mom and dad, they don't say how wonderful," notes Mariana. Lexi replies, "Who cares? I wish I had two moms throwing me a quinceañera. My parents refuse, and we're Latino."

Here, the interaction between Mariana and Lexi highlights a few things. First, Lexi finds the dressmaker's surprised reaction of Stef and Lena's kiss humorous. This disclosure reflects her acceptance of her best friend's two mothers. She also sees having two mothers as a difference that does not matter since *they* are throwing Mariana a quinceañera. However, Mariana's response reveals her frustration, viewing the gesture of "when people say *bueno*" (read: Spanish speaking Latinos) as condescending. She would rather not be treated differently for having two lesbian mothers, noting that nobody observes "how wonderful" it is for Lexi to have a father and a mother (as that is the norm). Mariana's remarks also hint at Latina culture being traditionally conservative (due to its Catholicism), but perhaps becoming more progressively tolerant of homosexuality and same-sex couples.

In fact, Mariana appears to be torn between maintaining her Latina roots and accepting her "nontraditional" family. Audiences learn that Mariana's desire to have a normal quinceañera is so strong that she is embarrassed to dance with her mothers. She decides to have Mike Foster, Brandon's biological father, not Stef and Lena, for the first dance—traditionally a father-daughter dance. When Stef asks Mike, he agrees and asks if she is okay with Mariana's request. "Well, apparently dancing with your two moms is not exactly *quince* tradition. So, just roll with the punches," Stef explains. Here, we see the tension of conforming to tradition (i.e., the ritual of the father-daughter dance) with the contemporary reality (i.e., having two mothers).

Later, however, Mariana realizes just how much Lena and Stef have sacrificed, both financially and emotionally, to give their daughter the party she wanted. She apologizes for being selfish and not dancing with them. The episode ends with the three of them playfully dancing together. Here, the happy ending reaffirms that a ritual like the father-daughter dance can be reinvented for two mothers and a daughter.

Another example of how *The Fosters* adheres to tradition (i.e., beliefs and practices that audiences are familiar with) while simultaneously broadening cultural convention is when Stef and Lena decide to officially get married. In "I Do,"[60] preparations for the wedding begin. In the kitchen, Stef's mother Sharon (Annie Potts) asks if they will be walking down the aisle. "Two women walking down the aisle? Really? Who's waiting at the other end? Why are we being given away? To whom are we being given to? It's kind of silly, don't you think?" chuckles Stef. "I don't know if it's silly, exactly, but yeah, it's not us, that's true," a hesitant Lena replies.

We see the different attitudes about the wedding ritual between Lena and Stef. While Lena seems to agree that the tradition of the bride walking down the aisle is "not us" (and therefore, wedding traditions can be adapted to fit them as a lesbian couple), Stef (who had been previously married to a man, Mike), appears to be a bit defensive of the custom in which the bride's father is to "give away" his daughter to the groom. She labels it "silly," especially as it will be "two women" engaging in the ritual. In many ways, her remarks reflect how traditional weddings are innately heteronormative—there is usually a bride and a groom, not two brides. As Chrys Ingraham asserts, "white weddings, while important in themselves, are a concentrated site for the operation and reproduction of organized heterosexuality."[61]

Later on in the episode, Lena and Stef get into a fight. "You've been downright hostile towards anything as ridiculous as walking down the aisle, or wearing a dress, or even writing vows," remarks Lena. She questions Stef's commitment in getting married due to her negative attitude towards these ceremonial acts. While they may be entrenched in heterosexual tradition, the wedding has symbolic meaning that goes beyond heterosexuality—it is a public affirmation of the loving relationship between them, which Stef soon comes to realize. She goes and confronts her father, telling him that she has been "really embarrassed about this wedding" because "this voice in my head keeps telling me it's not right what we're doing." She explains, however, it is not her voice, but *his* voice in her head—the conservative, religious, homophobic voice that she has internalized—and asks him not to attend the wedding.

Although Frank is absent, as with any traditional wedding, family and friends gather around in the decorated backyard. Both Lena and Stef's mothers are included in the ceremony, sharing words of wisdom and reflecting their support for their daughters' union. Lena's father, the Reverend Stewart Adams (Stephen Collins), officiates the wedding and explains that "now that the Supreme Court has finally seen fit to recognize that all people are entitled to equal protection under the law and have the right to the same challenges and triumphs, benefits and burdens, as everybody else. I'm very proud to welcome you here to the marriage of my daughter, Lena Elizabeth Adams, to

Stefanie Marie Foster."[62] Clearly, the show's depiction of a same-sex wedding ceremony becomes not only an opportunity to affirm and normalize Lena and Stef's relationship, it also becomes a site for political support of same-sex marriage and families. Reverend Adams's words that "all people are entitled to equal protection under the law" emphasizes the *legal* aspect of their marriage, that the traditional understanding of the institution of marriage should be extended to gay and lesbian couples.[63]

Normalizing Sexual Intimacy

The second major theme is that of *The Fosters* normalizing queer sexual intimacy. Throughout the series, we see Stef and Lena being affectionate with one another like any healthy monogamous couple. Audiences see them exchanging a quick kiss hello or goodbye. Nonetheless, *The Fosters* goes beyond the modest displays of affection, depicting sexually intimate scenes between Stef and Lena, a rarity with television shows and films that include lesbian couples[64] and gay parents.[65]

For example, at the end of "The Morning After,"[66] Lena and Stef are alone in the car, parked in their driveway. They apologize to one another for a disagreement they had that morning. Then, Stef tells Lena that she is "a hot saint" and kisses her. Lena says they can't fool around in the car because "the kids are right inside." Lena turns on some music and they passionately kiss. They jump into the backseat of the car. Stef's foot accidently kicks the horn, and they both giggle. While the episode ends without audiences "seeing" any sexual activity, only the insinuation that Stef and Lena will have a quick sexual rendezvous in the car, the scene normalizes lesbian sexuality and sex within a marriage—it can be loving, playful, and spontaneous. Their conversation also reflects this dialectical tension between motherhood (i.e., a non-sexual identity) and lesbian partner (i.e., a sexual identity). In fact, instead of having them in the house, a domestic space that symbolizes motherhood, Stef and Lena remain in the car, a symbolic site of teen sexual fantasy, an adventurous private space in which they could easily be "caught" by their children or a passing neighbor.

Additionally, sex and sexual intimacy are central to Stef and Lena's romantic relationship. In fact, as Riel Lise notes, "while *The Fosters* doesn't define its leading ladies by their sexuality, it does explore how their queerness affects their relationships and their views of the world."[67] Viewers are presented with a loving queer relationship with sexual chemistry between two women. In the opening scene of "The Honeymoon," we see Stef and Lena in bed, the warm morning light streaming across them. Their exposed shoulders suggest that they have consummated their marriage. "Good morning, Mrs.

Adams Foster," says Lena. "Good morning, Mrs. Adams Foster," replies Stef. They joke that they have decided that there will be no hyphen with their last name. "But how much do I love that we all have the same last names now," says Lena. "I love it, I love it, I love," kisses Stef on Lena's forehead. Lena spoons Stef and tells her that the only thing she has to do is kiss her, although Stef resists say she has morning breath. "Come here, woman," Lena seductively says. "I love it when you call me woman," Stef replies. She turns, and they embrace. Here, the show normalizes sexual desire between two women.[68]

However, there are challenges to Lena and Stef's relationship, as with any long-term committed relationship or marriage. In "Play," Lena and Stef decide to take a "babymoon" and leave the kids responsible for taking care of themselves for one night.[69] That evening in the hotel room, Lena and Stef are in bathrobes. Lena tells Stef that there is lesbian porn. "How does *Breakfast on Tiffany* sound?" asks Lena. "That sounds horrifying," chuckles Stef. Lena tells them it could "help us get into the mood." "Please. No bigger turn off than two straight girls in stilettos pretending for men, that they're hot for each other," Stef replies. Stef inadvertently dismisses Lena's appeal to engage in sexual intimacy by critiquing the lesbian porn. A hurt Lena goes and takes a bath. When Stef tries to smooth the situation over, Lena explains that she feels disconnected from her. "I want you to talk to me. About something other than money or the kids." Later, after apologies are exchanged, the two return to the hotel room. Stef seductively dances for Lena and takes her shirt off. They intensely kiss on the bed.

Rather than lesbian sexuality being stereotypically represented, perpetuating the heterosexual male fantasy,[70] or completely absent, *The Fosters* presents a much more complex and realistic representation of intimacy (emotional and sexual) within a lesbian relationship. In fact, as Melanie E. S. Kohnen notes, "the framing of intimacy as a significant component of a healthy relationship also sets *The Fosters* apart from other cable programs that use lesbian sexuality as 'edgy' or [as a] titillating element."[71] Queer sexuality is viewed as similar to heterosexuality. Sexual desire is linked to closeness and care, an expression of lust and love. Sexual intimacy is something that must be communicated and interpersonally negotiated.

While the series normalizes romantic and sexual intimacy in lesbian relationships, it also acknowledges the risks gay and lesbian couples face by engaging in public displays of affection. For example, in "The Morning After" episode, Lena discloses to Jude about issues of safety in a homophobic world after he is bullied for wearing nail polish.

> LENA: Sometimes when we're out in a new neighborhood or walking home late to our car, we won't hold hands.
> JUDE: Why?
> LENA: Some people out there are afraid of what's different, and sometimes they

want to hurt people like Stef and me. So, every time we're out and I want to hold Stef's hand, but I decide not to, I get mad. And mad at the people who might want to hurt us, but mad at myself, too, for not standing up to them. 'Cause the thing is if you're taught to hide what makes you different, you can end up feeling a lot of shame about who you are, and that's not okay. There is nothing wrong with you for wearing nail polish, just like there's nothing wrong with me for holding Stef's hand. What's wrong is the people out there who make us feel unsafe.[72]

Here, Lena communicates the lived reality for many same-sex couples and LGBT persons—that there is shame in hiding in the closet, anger of self-censorship, and fear of being attacked for their sexual orientation.[73] However, she also notes that even though they are "different," that there is "nothing wrong" with who they are or the desire to express their gender and sexual identities. Heterosexual audiences, then, are provided an opportunity to recognize the unique challenges same-sex couples face in public spaces and reflect on the taken-for-granted informal privilege of holding hands in public with their romantic partner.[74]

Managing Parental Roles

A third major theme is the negotiation of parental roles between Lena, Stef, and Mike. As Stef's ex-husband, Mike plays a role in the lives of the Foster family, both as biological father to Brandon as well as sporadically as a father figure to the other children. In addition, Mike and Stef work together as officers (and briefly as partners) for the San Diego police department. The relationship between Mike and Stef appears to be a close, although sometimes a contentious, one. Mike is present at Mariana's quinceañera and dances the father-daughter dance with her. He also attends Stef and Lena's wedding. When Lena and Stef search for a runaway Callie, it is Mike who steps in to take care of the children. He rushes around the kitchen, attempting to make breakfast and lunches to accommodate the diverse dietary needs of Jude, Jesus, Brandon, and Mariana. In season two, Mike and Stef chat at work about Lena's pregnancy. When Stef explains that she is stressed by having six children, he tells her, "You and Lena are unbelievable moms. I'm really happy for you guys." Here, he verbally discloses his support for them.[75]

At times, Mike becomes a father figure for Jesus. In the "House and Home"[76] episode, Jesus is playing basketball with Mike when he faints. Mike, Lena and Stef sit down to talk to Jesus about his ADHD medication and the fact that his fainting is a side effect of not taking the drugs. Mike brings up behavioral modifications and suggests wrestling as a beneficial alternative. At first, Jesus resists: "I'm not groping dudes." Mike corrects him, clarifying that wrestling is about "using strategic moves to take down your opponent."

Here, Mike reframes wrestling as a masculine activity, absent of homoerotic meanings (although he does not correct Jesus' implicitly homophobic comment). Of course, it is only after his mothers tentatively approve of the idea that Jesus can try out wrestling, but it is Mike who accompanies Jesus to wrestling practice, taking on a father-like role.

For the most part, the relationship between Mike and Stef (and Lena) is amicable; however, it is not without challenges. For example, throughout the series, Mike struggles with his alcohol addiction, sometimes letting down his son Brandon with his inconsistent behavior. And when Mike is present in the parental decision-making process, he does not always see eye-to-eye with Stef and Lena. This tension of co-parenting Brandon is seen throughout the series, particularly when Brandon helps Callie rescue her brother Jude from a violent foster father but puts everyone's lives at risk.[77] The mothers' decision not to punish Brandon leads Stef and Lena into a heated argument with Mike. Here, parenting styles and personalities clash. When Lena tries to explain that they are not grounding him, Mike gets irritated. "With all due respect, what you think really isn't my concern," snaps Mike. "Hey, wait a minute, Mike. Lena is his parent too," replies Stef, defending her wife. "Respectfully, that may be true. But I'm his father and you're his mother," Mike responds, reinforcing biological kinship and heteronormativity. "That's fine. I'm not his mother. I'm not his father. Mike is right," says an exasperated Lena who, since she has not yet married Stef, has no legal parental rights to Brandon. "No, he is not," says Stef, reiterating that parental roles and responsibilities go beyond blood ties. However, when asked about Brandon's punishment, Stef is caught between Lena and Mike.

Later, Lena confronts Stef for not taking her side and making her feel like "a stepmother." Stef tries to explain that Mike was upset that he was not included in the parental decision making. "You humiliated me in front of him," says Lena, feeling excluded from the parental process. Stef repeats that "Brandon has three parents, period," affirming Lena's role as his mother. Stef also notes the unique challenges they face with the non-normative structure of three parents by saying, "We're all trying to do our best to figure this out." Ultimately, *The Fosters* does just that—portraying Lena and Stef negotiating their roles as parents, whether they be foster mothers, adoptive mothers, stepmothers, or biological ones.

This theme of negotiating parental boundaries among Stef, Lena, and Mike is seen again after Brandon's hand is severally injured in an attack. In "Take Me Out," Brandon announces he wants to have the surgery on his hand in hopes of getting all of his feeling back.[78] Stef goes to see Mike to make sure they are on the same page. Unfortunately, when Stef explains to Brandon that she and Mike have decided against it, Lena questions her role in the decision-making process. Stef states that she and Mike already discussed the risks involved. "And you made your decision. Got it," an angry Lena replies. Again, Lena seems to be excluded from the parental decision-making process since Brandon is neither her biological son nor an adopted one. Later, in private, Lena tells Brandon she thinks it is his decision but that "I think Mom and your Dad are wrong on this one." Stef overhears their conversation and confronts Lena. "Did I really just hear you tell our son that you think he should get the surgery?" She tells Lena she upset that they are not a "unified front" and that Lena has undermined her authority "with our son." Here, the use the phrase of "our son" by Stef seems to symbolically legitimize Lena as a parent; however, Lena questions Stef's use of the possessive pronoun with "Our son? Or your son?" Clearly, she is angry at not being consulted about the situation. Lena also seems to recognize their difference in mothering styles—she trusts Brandon to make his own medical decisions because he is old enough, while Stef views her responsibility as a parent to make decisions for him.

In the end, Lena and Stef talk to Brandon. "Your Mom and I have been talking," Stef starts off and affectionately grabs Lena's arm, nonverbally gesturing a unified front. By labeling Lena as "your Mom," she verbally validates Lena's parental role. Stef then admits she just wants to protect Brandon but does not know what the right choice is for him. The episode concludes with Brandon deciding that he will hold off on the surgery.

Reframing Discourses of Family

The Fosters attempts to resist the dominant heteronormative image of family (i.e., husband and wife with biological children) by depicting family as a married, interracial lesbian couple with biological, foster, and adopted children. However, the show does not then display a lesbian headed household as radically different from the traditional family. Despite the fact that Stef's lesbian sexuality is seen as breaking the normative heterosexual family unit of husband, wife, and child when she divorces Mike, family is "reproduced" when she enters into a monogamous relationship with Lena. Family is still defined by monogamy and having children rather than by just the couple unit.

The series also appeals to the universal—arguing that the most important characteristic in defining family is love. For example, in the theme song lyrics, we hear about the importance of love, while emphasizing it is not where one is born, but finding the place where one wants to be and is wanted. Family is viewed not as biological but rather as a place of love, safety, and belonging, a home where one will find a family and no longer remain alone. Additionally, images of the domestic sphere go along with the song—a montage of a family chore chart, mail on the stairs, dishes in the sink, pillows and beds, children's names and heights marked on the wall, breakfast foods, picture frames, and a refrigerator door cluttered with photos and recipes. Thus, home is where the family (and the heart) is. Yet, family is not taken for granted; rather, time and time again, characters verbalize definitions of familial relationships. For example, after Callie questions the legitimacy of adopted children as "real" kin, Stef reassures Jesus and Mariana that "you two are every bit as much our kids as Brandon is."[79] After finding out that Mariana secretly met with Ana (Alexandra Barreto), Mariana's birth mother, Lena talks to her about family: "I know it must be hard to understand how your birth mom could have chosen drugs over you and your brother. I don't understand it. All I know is we chose you. And you chose us. DNA doesn't make a family. Love does."[80] At Stef's father funeral, Lena thanks Callie, who currently resides at the Girls United group home, for coming and tells her that "you know you're always going to be a member of this family no matter what." Later in the episode, when Callie tells Lena's mother, Dana Adams (Lorraine Toussaint), that she is sixteen and does not need to be adopted, Dana responds with "Family isn't just until you're 18, you know. You need them your whole life."[81] Time and time again, audiences not only *see* diverse familial relationships but also *hear* family members talk about what family means.

Family kinship, then, is defined not just by blood, but by love and "by means of deep emotional and symbolic ties with home, society, and tradition."[82] In fact, the show seems to resist the idea that biological parents are inherently superior or that children should necessarily be with their birth parents. Jesus and Mariana's birth mother is a drug addict who first contacts Mariana for money and then later attempts to blackmail the family. Donald Jacob (Jamie McShane), Jude's biological father, was imprisoned for manslaughter (for a drunk driving accident that killed Colleen, their mother) and gives up his parental rights so Jude can be adopted, noting that he is his biological "dad" but not "a parent," defined as someone who has the financial means and skills necessary to care for Jude.[83]

The Adams Foster family depicts a variety of kinship relationships that are often absent or ignored on television, particularly with regard to foster children and adopted children. *The Fosters* emphasizes the challenges of the foster system and the messiness of the adoption process—both which appear to have a birth parent bias. For example, when Lena and Stef petition the court to adopt Jude and Callie, viewers learn that the brother and sister do not share the same biological father. Callie's adoption is denied by Judge Ringer, who acknowledges Donald as the "presumptive father"[84] but determines that he is not the paternal father of Callie based upon the birth certificate (an official legal document). Therefore, Donald does not have legal status to "terminate parental rights." It is Callie's biological father, Robert Quinn (Kerr Smith), who has the authority to allow her to be adopted by Lena and Stef.

A few episodes later,[85] Callie discloses to Jude the existence of her half-sister, Sophia (Bailee Madison), Robert's daughter. She explains that Donald is "our dad" but that Robert is the man she is genetically linked to. When Jude states that he is like Sophia, Callie's half-sister, she corrects him: "You are my brother. Period." Jude is accurate that he is Callie's half-brother since they biologically share the same mother, but Callie's response is way to affirm that their relationship has not changed. Callie emphasizes that while they may have different fathers, their sibling bond is deeper than DNA.

At the same time *The Fosters* appears to embrace these "alternative" familial relationships, the show also problematizes these kinship ties. One of the major narrative arcs in the first season is the romantic relationship between Callie and Brandon. This is complicated by the fact that they are foster sister and brother and therefore are not allowed to be romantically involved. When Brandon confesses to Callie that he is in love with her, she responds:

> CALLIE: Brandon, I'm going to be your sister.
> BRANDON: We're not related. We don't share the same blood. There's nothing wrong or illegal about that.
> CALLIE: So if you're not really my brother, then Stef and Lena won't really be my mothers.
> BRANDON: Not, it's not the same.
> CALLIE: It's exactly the same. I need a family. Not a piece of paper. If that's what adoption is, then what's the point?[86]

For Callie, to be romantically involved with Brandon would be symbolically incestuous, particularly after she has been adopted. In the end, Callie's dream for a family is too great. She decides to sacrifice a potential relationship with Brandon so that she can be a "real" daughter with "real" mothers. Through Callie, audiences hear the legitimization of adoptive familial ties—they are real.[87]

Finally, along with adoptive and foster relationships, *The Fosters* provides audiences with a look at artificial insemination with the storyline of Lena desiring to have a baby. This storyline speaks to how many gay and lesbian couples struggle with reproductive technologies (e.g., egg donation, sperm donation, in vitro fertilization, surrogacy). It also resists dominant discourses about family-making given that "alternative insemination, in particular, is a technique for acquiring children that challenges conventional understandings of biological offspring" as a child is viewed as a "natural" product between a man and woman.[88]

When Lena tells Stef about wanting to get pregnant through insemination,

Stef is reluctant as they already have five teenagers. However, as Stef has already had a biological child (Brandon), she decides Lena should be allowed the opportunity to do the same. At first, they decide they want an anonymous donor. They peruse sperm donors on the computer. When Lena selects a donor who is African American, Stef responds that she wishes that since they cannot have a biological baby together, that the child "has more of my characteristics." Lena explains she is half white, and if the donor is white, the baby might not look like her at all.[89]

At school, Lena gets the idea that the good-looking English professor, Timothy (Jay Ali), might be a potential donor. After an initial awkward dinner, Lena and Stef privately discuss his positive attributes (attractive, good with students, Rhodes Scholar, published a novel, plays the sitar). "And he's kind?" Stef asks. "Very," replies Lena. Stef and Lena hold hands, signifying that they have agreed that Timothy is their donor choice. Timothy, also, is perhaps a racial compromise; he is neither black like Lena nor white like Stef, but Indian.[90]

However, things become complicated when Timothy backs out of signing the donor contract which would waive his parental rights. When he finally does accept the contract, Stef discusses how perhaps when the child is born, they could include him "not as a father, but as a donor and a friend." Unfortunately, after all the parental drama, Lena ends up having to terminate the pregnancy due to a severe case of pre-eclampsia.[91] The choice to have a late-term abortion is heartbreaking. Lena cannot risk her life for a premature child that may not survive. She chooses the needs of her children over her own.

A crucial scene occurs when Dana, Lena's mother, comes to visit her in the hospital. Lena explains, "I know this sounds selfish, but everyone in the house is connected to somebody by blood. I know that you don't need that to be a family. But sometimes I'm afraid they won't love me as much as I love them because I'm not their biological mom, that they won't have that to tether them to me."[92] Here, Lena speaks to her anxiety of the tentativeness of non-biological family ties. What if these adopted children she loves so much leave her? Dana responds that all mothers (biological or not) have these fears, but that they all must always must love more. "Love will always be a stronger bond than blood," she wisely explains, echoing the theme of the television show.

Conclusion

In this essay, we critically examined the relationship between Lena and Stef, the mothers of the show, and the ways in which the teen drama *The Fos-*

ters both reinforces and resists dominant discourses of parenting, sexuality, and gender within the genre of family programming. By centering the family with a lesbian couple, *The Fosters* legitimizes and normalizes gay parenting, extending the definition of family to be inclusive of "alternative" family forms. As Renee Fabian notes, "By creating characters who wrestle with everyday concerns, these people are easy for audiences to identify with, regardless of sexual orientation or gender; a parent is a parent."[93]

By engaging in a close textual analysis of the first two seasons of *The Fosters*, we identified four major themes. First, we found that the show demonstrates that traditional rituals are an important aspect of family life but are re-envisioned in contemporary ways that are inclusive to gay and lesbian couples. Second, sexual intimacy is depicted as an important aspect of romantic lesbian relationships and is communicatively negotiated within the demands of marriage and motherhood. Third, while the series positions family life within lesbian couplehood, parental roles are continuously negotiated by Stef, Lena, and Mike within the non-normative structure of three parents. Finally, *The Fosters* reframes discourses of family to include diverse parent-child formations (fostered, adopted, biological) but continues to maintain family as centered on childrearing.

Media scholars interested in LGBT issues must continue to examine critically how television, particularly teen and family dramas, represents queer individuals and gay- and lesbian-headed families. As Laura Oswald notes, "as the number of family configurations flourishes, the very notion of family has been broken down into a plurality of meanings."[94] Television programs have become a space for depicting a diversity of family formations (e.g., *Modern Family, Fresh Off the Boat* [ABC, 2015–present], *Black-ish* [ABC, 2014–present]); nonetheless, an exploration of how these fictional representations reshape notions of family into these "plurality of meanings" is still required.

Additionally, some scholars have noted the lack of mediated messages associated with healthy non-heterosexual relationships on television,[95] while other scholars have observed that teen television programming often embraces non-heterosexual identities.[96] Therefore, more research is needed to explore *how* queer characters and relationships are represented and *how* they shape audiences' views about LGBT issues (e.g., same-sex marriage, public bathroom policies, bullying, and teen suicide). While increased visibility of LGBT characters, gay and lesbian parents, and same-sex parenting on television suggests a greater cultural acceptance of queer identities, normalizing gay and lesbian relationships and queer families as "just like everyone else" can be problematic as it reinscribes queerness into larger heteronormative structures. That is, there is an erasure of unique stressors and challenges LGBT individuals and families face from their heterosexual counterparts.

With LGBT rights threatened by discriminatory policies and laws across the United States (e.g., allowing business to deny LGBT individuals services due to religious reasons, forcing transgender individuals to use restrooms based sex on their birth certificate rather than their chosen gender identity), television becomes a powerful tool for educating audiences about current cultural issues and LGBT advocacy. Specifically, legal struggles continue to persist about who is a parent and what constitutes parenthood. For example, in 2016, Indiana federal judge Tanya Walton Pratt ruled that both same-sex parents should be listed on their children's birth certificates, citing that it is unconstitutional to list both parents only when a mother and a father are involved.[97] The state of Indiana is appealing this ruling, reflecting the still-controversial nature of same-sex couplehood and the bias towards biological, heteronormative constructions of parenthood. Shows like *The Fosters*, then, can provide audiences with more realistic depictions that humanize lesbians and lesbian-headed families, that portray the difficulties of the foster system and adoption, and that potentially help to broaden definitions of family. In the end, the message on *The Fosters* is clear—family, in all of its many forms, is about love.

Notes

1. Bradley Bredeweg and Peter Paige, "Pilot," *The Fosters*, season 1, episode 1, ABC Family, directed by Timothy Busfield, aired June 3, 2013.

2. Montgomery Jones, "Must-See TV: The Fosters," *Spark Movement,* September 4, 2013, http://www.sparkmovement.org/2013/09/04/must-see-tv-the-fosters/.

3. Mike Hale, "Gay, Hispanic, Adoptive: A Triple-Threat Family," *The New York Times,* June 2, 2013, http://www.nytimes.com/2013/06/03/arts/television/the-fosters-on-abc-family.html.

4. Genevieve Valentine, "*The Fosters* Is More Than the Sum of Its Very Special Episodes," *A.V. Club*, January 13, 2014, http://www.avclub.com/review/the-fosters-is-more-than-the-sum-of-its-very-speci-200803; Todd VanDerWerff, "The Fosters," *A.V. Club*, June 3, 2013, http://www.avclub.com/tvclub/emthe-fostersem-98519; Yvonne Villarreal, "'The Fosters,' an ABC Family Drama, Pushes the Envelope Further," *Los Angeles Times*, June 19, 2013, http://articles.latimes.com/2013/jun/19/entertainment/la-et-st-the-fosters-20130619.

5. Mike Hale, "A Family Drama, and a Balancing Act," *The New York Times*, January 12, 2014, https://www.nytimes.com/2014/01/13/arts/television/the-fosters-on-abc-family-about-a-same-sex-couple.html?_r=0.

6. Riel Lise, "Hooray for 'The Fosters,' The Show About a Dynamic Family with Two Moms," *Bitch Media*, January 30, 2015, https://bitchmedia.org/post/hooray-for-the-fosters—the-show-about-dynamic-family-with-two-moms.

7. Sarah Caldwell, "The Fosters," *Entertainment Weekly*, June 4, 2013, http://ew.com/article/2013/06/04/fosters/.

8. Melanie E. S. Kohnen, "Cultural Diversity in Brand Management in Cable Television," *Media Industries* 2, no. 2 (2015), http://www.mediaindustriesjournal.org/index.php/mij/article/view/124/181#sdendnote6anc; Willa Paskin, "'The Fosters' Could Be This Summer's Answer to 'The OC,'" *Salon*, June 3, 2013, http://www.salon.com/2013/06/03/the_fosters_is_solid_summer_television/.

9. Tim Gray, "Jennifer Lopez on TV Diversity: 'A Show's Characters Need to Reflect Today's Society,'" *Variety*, June 29, 2015, http://variety.com/2015/tv/news/jennifer-lopez-the-fosters-marriage-equality-lgbt-1201529644/.

10. Kohnen, "Cultural Diversity."

11. Alysia Abbott, "TV's Disappointing Gay Dads," *The Atlantic*, October 31, 2012, http://www.theatlantic.com/sexes/archive/2012/10/tvs-disappointing-gay-dads/264134/#slide1; Spencer Kornhaber, "The *Modern Family* Effect: Pop Culture's Role in the Gay-Marriage Revolution," *The Atlantic*, June 26, 2015, https://www.theatlantic.com/entertainment/archive/2015/06/gay-marriage-legalized-modern-family-pop-culture/397013/.

12. Rich Valenza, "Bradley Bredeweg, Executive Producer, Discusses ABC Family's 'The Fosters,'" *Huffington Post*, January 10, 2014, http://www.huffingtonpost.com/2014/01/10/bradley-bredeweg-the-fosters_n_4569014.html?utm_hp_ref=mostpopular.

13. Renee Fabian, "How ABC Family Is Getting It Right: 'The Fosters' and 'Pretty Little Liars,'" GLAAD, December 17, 2013, http://www.glaad.org/blog/how-abc-family-getting-it-right-fosters-and-pretty-little-liars.

14. James R. Keller, *Queer (Un)Friendly Film and Television* (Jefferson, NC: McFarland, 2002); Jane Campbell and Theresa Carilli, eds., *Queer Media Images: LGBT Perspectives* (Lanham, MD: Lexington Books, 2013); Fred Fejes and Kevin Petrich, "Invisibility, Homophobia and Heterosexism: Lesbians, Gays and the Media," *Critical Studies in Mass Communication* 10, no. 3 (1993): 396–422.

15. Maria San Fillippo, *The B Word: Bisexuality in Contemporary Film and Television* (Bloomington: Indiana University Press, 2013), 4.

16. Kathleen Battles and Wendy Hilton-Morrow, "Gay Characters in Conventional Spaces: *Will and Grace* and the Situation Comedy Genre," *Critical Studies in Media Communication* 19, no. 1 (2002): 87–105; Amber B. Raley and Jennifer L. Lucas, "Stereotype or Success? Prime-Time Television's Portrayals of Gay Male, Lesbian, and Bisexual Characters," *Journal of Homosexuality* 51, no. 2 (2006): 19–38.

17. Vito Russo, *The Celluloid Closet: Homosexuality in the Movies*, rev. ed. (New York: Harper & Row, 1987), 122.

18. Frederick Dhaenens, "The Fantastic Queer: Reading Gay Representations in *Torchwood* and *True Blood* as Articulations of Queer Resistance," *Critical Studies in Media Communication* 30, no. 2 (2013): 110–111.

19. Dustin Bradley Goltz, *Queer Temporalities in Gay Male Representation: Tragedy, Normativity, and Futurity* (New York: Routledge, 2010), 17.

20. *Ibid.*, 21.

21. Ann M. Ciasullo, "Making Her (In)visible: Cultural Representations of Lesbianism and the Lesbian Body in the 1990s," *Feminist Studies* 27, no. 3 (2001): 577–608.

22. C. Lee Harrington, "Lesbian(s) on Daytime Television: The Bianca Narrative on *All My Children*," *Feminist Media Studies* 3, no. 2 (2003): 216.

23. Bonnie Dow, "Ellen, Television, and The Politics of Gay and Lesbian Visibility," *Critical Studies in Media Communication* 18, no. 2 (2010): 137.

24. Ciasullo, "Making Her (In)visible," 578.

25. Kay Seibler, *Learning Queer Identity in the Digital Age* (New York: Palgrave Macmillan, 2016), 87–90.

26. Katerina Symes, "Orange Is the New Black: The Popularization of Lesbian Sexuality and Heterosexual Modes of Viewing," *Feminist Media Studies* 17, no. 1 (2017): 29.

27. Pei-Wen Lee and Michaela D. E. Meyer, "'We All Have Feelings for Our Girlfriends': Progressive (?) Representations of Lesbian Lives on *The L Word*," *Sexuality & Culture* 14, no. 3 (2010): 245.

28. Michaela D. E. Meyer, "'It's me. I'm it': Defining Adolescent Sexual Identity Through Relational Dialectics in *Dawson's Creek*," *Communication Quarterly* 51, no. 3 (2003): 262–276.

29. Michaela D. E. Meyer, "'I'm Just Trying to Find My Way Like Most Kids': Bisexuality, Adolescence and the Drama of *One Tree Hill*," *Sexuality and Culture* 13, no. 4 (2009): 237–251.

30. Michaela D. E. Meyer and Megan M. Wood, "Sexuality and Teen Television: Emerging Adults Respond to Representations of Queer Identity on *Glee*," *Sexuality & Culture* 17, no. 3 (2013): 435.

31. Jane D. Brown, "Mass Media Influences on Sexuality," *The Journal of Sex Research*

39, no. 1 (2002): 42–45; Alfred P. Kielwasser and Michelle A. Wolf, "Mainstream Television, Adolescent Homosexuality, and Significant Silence," *Critical Studies in Mass Communication* 9, no. 4 (1992): 350–373.

32.　Rachel E. Silverman, "Family Perfection: The Queer Family as Perspective by Incongruity on *Will & Grace* and *Queer as Folk*," in *Queer Media Images: LGBT Perspectives*, ed. Jane Campbell and Theresa Carilli (Lanham, MD: Lexington Books, 2013); Jessica Murrell and Hannah Stark, "Allegories of Queer Love: Quality Television and the Reimagining of the American Family," in *Queer Love in Film and Television: Critical Essays*, ed. by Pamela Demory and Christopher Pullen (New York: Palgrave Macmillan, 2013).

33.　Sophia Laubie, "Gay Parents on TV: Why The 'New Normal' Is No Longer Just the Nuclear Family," *Huffington Post,* December 7, 2013, http://www.huffingtonpost.com/2013/12/07/gay-parents-tv_n_4402297.html.

34.　Peter C. Kunze, "Family Guys: Same-Sex Parenting and Masculinity in *Modern Family*," in *Queer Love in Film and Television: Critical Essays*, ed. Pamela Demory and Christopher Pullen (New York: Palgrave Macmillan, 2013), 111.

35.　Steven E. Doran, "Housebroken: Homodomesticity and the Normalization of Queerness in *Modern Family*," in *Queer Love in Film and Television: Critical Essays*, ed. Pamela Demory and Christopher Pullen (New York: Palgrave Macmillan, 2013), 101.

36.　Michael Warner, *The Trouble with Normal: Sex, Politics, and the Ethics of Queer Life* (New York: The Free Press, 1999), 60.

37.　Andre Cavalcante, "Anxious Displacements: The Representation of Gay Parenting on Modern Family and The New Normal and the Management of Cultural Anxiety," *Television & New Media* 16, no. 5 (2015): 460.

38.　Suzanna Danuta Walters, "The Kids Are All Right, but Lesbians Aren't: Queer Kinship in U.S. Culture," *Sexualities* 15, no. 8 (2012): 920.

39.　Stephanie Coontz, *The Way We Never Were: American Families and the Nostalgia Trap* (New York: Basic Books, 1992), 183.

40.　Gary J. Gates, "LGBT Parenting in the United States," The Williams Institute, UCLA School of Law, February 2013, http://williamsinstitute.law.ucla.edu/wp-content/uploads/LGBT-Parenting.

41.　Tor Folgero, "Queer Nuclear Families? Reproducing and Transgressing Heteronormativity," *Journal of Homosexuality* 54, no. 1/2 (2008): 125.

42.　Julia Erhart, "Donor Conception in Lesbian and Non-Lesbian Film and Television Families," in *Queer Love in Film and Television: Critical Essays*, ed. Pamela Demory and Christopher Pullen (New York: Palgrave Macmillan, 2013), 83–93.

43.　Nancy J. Mezey, *New Choices, New Families: How Lesbians Decide About Motherhood* (Baltimore: Johns Hopkins University Press, 2008).

44.　Ellen Lewin, *Lesbian Mothers: Accounts of Gender in American Culture* (Ithaca: Cornell University Press, 1993).

45.　Valerie Lehr, *Queer Family Values: Debunking the Myth of the Nuclear Family* (Philadelphia: Temple University Press, 1999).

46.　Liza Mundy, "The Gay Guide to Wedded Bliss," *The Atlantic*, June 2013, https://www.theatlantic.com/magazine/archive/2013/06/the-gay-guide-to-wedded-bliss/309317/.

47.　Warner, *The Trouble with Normal*, 127.

48.　James Jasinski, *Sourcebook on Rhetoric: Key Concepts in Contemporary Rhetorical Studies* (Thousand Oaks, CA: Sage, 2001), 93.

49.　Jennifer C. Dunn and Stephanie L. Young, *Pursuing Popular Culture: Methods for Researching the Everyday* (Dubuque: Kendall Hunt, 2017), 21.

50.　Barry S. Brummet, *Techniques of a Close Reading* (Thousand Oaks, CA: Sage, 2010), 101.

51.　Dunn and Young, *Pursuing Popular Culture*, 49.

52.　Brian L. Ott and Robert L. Mack, *Critical Media Studies*, 4th ed. (Malden: John Wiley & Sons, 2014), 196.

53.　Michaela D. E. Meyer, "'It's me. I'm it,'" 264.

54.　Adam Liptak, "Supreme Court Ruling Makes Same-Sex Marriage a Right Nationwide," *The New York Times*, June 26, 2015, https://www.nytimes.com/2015/06/27/us/supreme-court-same-sex-marriage.html.

55. Steven J. Wolin and Linda A. Bennett, "Family Rituals," *Family Process* 23, no. 3 (1984): 401.

56. Ann Swidler, "Culture in Action: Symbols and Strategies," *American Sociological Review* 51, no. 2 (1986): 279.

57. Joanna Johnson, "Quinceañera," *The Fosters*, season 1, episode 4, ABC Family, directed by Joanna Kerns, aired June 24, 2013.

58. While Mariana's Latina identity is brought up throughout the series, her specific ethnicity is not. Audiences may assume she is Mexican American as her birth mother's last name is Gutierrez, but this is unclear.

59. Julia Alvarez, *Once Upon a Quinceañera: Coming of Age in the USA* (New York: Plume, 2007), 2.

60. Bradley Bredeweg and Peter Paige, "I Do," *The Fosters*, season 1, episode 10, ABC Family, directed by James Haymen, aired August 5, 2013.

61. Chrys Ingraham, *White Weddings: Romancing Heterosexuality in Popular Culture* (New York: Routledge, 2008), 3.

62. Bredeweg and Paige, "I Do."

63. As *The Fosters* is set in California, it should be noted that when Reverend Adams uses "Supreme Court," he may be referring to the California Supreme Court that legalized same-sex marriage in 2008 or the U.S. Supreme Court case *Hollingsworth v. Perry* (2013) that overturned Proposition 8 and legalized same-sex marriage in California. However, it would not be until 2015 with *Obergefell v. Hodges* that the U.S. Supreme Court would federally legalize same-sex marriage.

64. Walters, "The Kids Are All Right," 921–922; Sheena Howard, "*The Kids Are All Right*: A Mediated Ritual Narrative," *Women & Language* 36, no. 2 (2013): 81–87.

65. Calvancante, "Anxious Displacements," 465.

66. Paul Sciarrotta, "The Morning After," *The Fosters*, season 1, episode 5, ABC Family, directed by Bethany Rooney, aired July 1, 2013.

67. Riel Lise, "Hooray for 'The Fosters,' The Show About a Dynamic Family with Two Moms," *Bitch Media*, January 30, 2015, https://bitchmedia.org/post/hooray-for-the-fosters-the-show-about-dynamic-family-with-two-moms.

68. Peter Paige and Brad Bredeweg, "The Honeymoon," *The Fosters*, season 1, episode 11, ABC Family, directed by James Hayman, aired January 13, 2014.

69. Thomas Higgins, "Play," *The Fosters,* season 2, episode 3, directed by Martha Mitchell, aired June 30, 2014 (Burbank, CA: Freeform).

70. Ciasullo, "Making Her (In)visible," 578.

71. Kohnen, "Cultural Diversity."

72. Sciarrotta, "The Morning After."

73. Amy Steinbugler, "Visibility as Privilege and Danger: Heterosexual and Same-Sex Interracial Intimacy in the 21st Century," *Sexualities* 8, no. 4 (2005): 425–443.

74. Long Doan, Annalise Loehr, and Lisa R. Miller, "Formal Rights and Informal Privileges for Same-Sex Couples: Evidence from a National Survey Experiment," *American Sociological Review* 79, no. 6 (2014): 1175.

75. While they have shared many life dramas, the friendly relationship between Stef and Mike is perhaps due to the fact that they have been divorced for over ten years, giving them time to adjust to their roles.

76. Johanna Johnson, "House and Home," *The Fosters*, season 1, episode 12, ABC Family, directed by Martha Mitchell, aired January 20, 2014.

77. Bradley Bredeweg and Peter Paige, "Consequently," *The Fosters*, season 1, episode 2, ABC Family, directed by Timothy Busfield, aired June 10, 2013.

78. Megan Lynn and Wade Solomon, "Take Me Out," *The Fosters*, season 2, episode 2, ABC Family, directed by Elodie Keene, aired June 23, 2014.

79. Bredeweg and Paige, "Pilot."

80. Joanna Johnson, "Vigil," *The Fosters*, season 1, episode 9, directed by David Paymer, aired July 29, 2013 (Burbank, CA: Freeform).

81. Tamara P. Carter, "Padre," *The Fosters,* season 1, episode 15, ABC Family, directed by Millicent Shelton, aired February 10, 2014.

82. Laura Oswald, "Branding the American Family: A Strategic Study of the Culture, Composition, and Consumer Behavior of Families in the New Millennium," *Journal of Popular Culture* 37, no. 2 (2003): 327.

83. Marissa Jo Cerar and Zoila Amelia Galeano, "Don't Let Go," season 1, episode 19, ABC Family, directed by Zetna Fuentes, aired March 10, 2014.

84. Bradley Bredeweg and Peter Paige, "Things Unknown," season 2, episode 1, ABC Family, directed by Normal Buckley, aired June 16, 2014.

85. Kathleen McGhee-Anderson, "Say Something," season 2, episode 4, ABC Family, directed by Zetna Fuentes, aired July 7, 2014.

86. Joanna Johnson, "Metropolis," season 1, episode 20, ABC Family, directed by Martha Mitchell, aired March 17, 2014.

87. Later on in the series, Callie and Brandon continue to struggle with their romantic feelings. They end up exploring a romantic relationship. However, it is always constrained within the tension of familial relationships and the risk of Callie not being adopted.

88. Kath Weston, *Families We Choose: Lesbians, Gays, Kinship* (New York: Columbia University Press, 1991), 169.

89. Megan Lynn and Wade Solomon, "Kids in the Hall," season 1, episode 17, ABC Family, directed by Michael Grossman, aired February 24, 2014.

90. Interestingly, Timothy is never given a last name. Based upon his physical features, audiences may see him racially ambiguous character, perhaps Middle Eastern. He also has a British accent, suggesting he is British Indian. Finally, it is mentioned that he plays the sitar, a traditional Indian stringed instrument.

91. Erin Gloria Ryan, "ABC Family's *The Fosters* Has the Balls to Tackle Late Term Abortion," *Jezebel*, July 23, 2014, http://jezebel.com/abc-familys-the-fosters-has-the-balls-to-tackle-late-te-1609572961.

92. Joanna Johnson, "Mother," season 2, episode 6, ABC Family, directed by Lee Rose, aired July 21, 2014.

93. Renee Fabian, "How ABC Family Is Getting It Right."

94. Oswald, "Branding the American Family," 311.

95. Deborah A. Fisher, Douglas L. Hill, Joel W. Grube, and Enid L. Gruber, "Gay, Lesbian, and Bisexual Content on Television: A Quantitative Analysis Across Two Seasons," *Journal of Homosexuality* 52, no. 3–4 (2007): 167–188.

96. Valerie Wee, "Teen Television and the WB Television Network," in *Teen Television: Essays on Programming and Fandom*, ed. Sharon Marie Ross and Louisa Ellen Stein (Jefferson, NC: McFarland, 2008), 43–60.

97. Stephanie Wang, "State Appeals Ruling on Parental Rights for Same-Sex Couples," *Indianapolis Star*, January 31, 2007, http://www.indystar.com/story/news/politics/2017/01/31/state-appeals-ruling-parental-rights-same-sex-couples-birth-certificate-lawsuit/97252040/.

"Puerto Rican and redheaded!?"

Constructions of Race, Ethnicity and Identity on Switched at Birth

Donica O'Malley

Daphne Vasquez (Katie Leclerc), a white, redheaded college freshman, desperately begs her professor to let her into an over-enrolled chemistry course. Before deciding, Professor Marillo (Bess Armstrong) asks Daphne, "tell me something about yourself." Daphne replies that she is Puerto Rican and grew up in East Riverside (a poor, predominately Latinx neighborhood in metro Kansas City, Missouri). Professor Marillo is stunned—"Puerto Rican and redheaded!?" she asks, incredulously. She launches into an explanation of the genetics of red hair, while expressing amazement at Daphne's phenotypical appearance, given her ethnicity. She agrees to accept Daphne, who lost her hearing from meningitis at age 3, provided she can find an American Sign Language (ASL) interpreter who knows chemistry. Daphne gleefully accepts. She does not, however, explain to Professor Marillo that her red hair was inherited from her white biological parents, nor that the reason she grew up in a Puerto Rican household in East Riverside was that she was switched with another baby, Bay Kennish (Vanessa Marano), in the hospital after birth.[1]

This essay considers scenes such as this one from Freeform TV's popular show *Switched at Birth* (2011–2017). In the show, characters often both explicitly and implicitly debate whether social categories, such as race and ethnicity, are biological or cultural. They also define and redefine what family means to them, inside and outside of these categories. Within health communication and disability studies scholarship, *Switched at Birth* has been praised for its representations of Deaf culture, and specifically for its complex d/Deaf young

139

adult and teenager characters.[2] The show has also been lauded for its nuanced depictions of controversial subjects, such as ineffective college campus sexual assault policies.[3] There have, of course, been critics as well, but the overall response to the show has been positive.[4] As of yet, however, no one has undertaken an analysis of the complicated racial and ethnic representations on the show. In this essay, I argue that although *Switched at Birth* attempts to construct an understanding of family outside of the boundaries of genetic relationships, the contexts in which it does so are special cases; the default understanding of family is biological. Additionally, the show's pervading focus on DNA presents racial and ethnic categories as scientifically predetermined, inheritable, and discrete. Ultimately, *Switched at Birth* suggests that Bay, Daphne, and their families cannot escape their biological identities—which at times are conflated with their racial and ethnic identities—for better or for worse.

Switched at Birth premiered in 2011, when Freeform TV was still known as ABC Family. *Switched at Birth* fit well with the prior network's tagline of "A New Kind of Family." The "new family" created by the switch is multi-ethnic, multi-generational, and includes both hearing and Deaf characters. Building upon the work of Melanie E.S. Kohnen, I ask whether the diversity portrayed on the ABC Family/Freeform TV actually constitutes what Mary Beltrán has referred to as "meaningful diversity," or diversity in which characters of color are fully realized people, the writers and producers are knowledgeable about their lives, the diversity of the cast appears natural, and the production team takes advantage of the diversity of the story's setting or subject matter.[5] Responding to Beltrán's work, Kohnen theorizes that ABC Family/Freeform TV has cultivated a particular type of diversity, which she refers to as "branded diversity," or "the inclusion of cultural diversity in television programming that is motivated by and contributes to a channel's branding strategies."[6] Importantly, Kohnen does not believe that such representations are automatically valueless, or to build off of Beltrán's term, "unmeaningful," due to their corporate motivations. Rather, she suggests that scholars should consider both the possibilities and constraints of these kinds of representations.[7] Therefore, throughout this essay, I follow Kohnen's suggestion, and in addition to critiquing the ways in which the show treats race and ethnicity, I also attempt to acknowledge instances where *Switched at Birth* does offer meaningfully diverse and complex representations through its characters.

Although praise of the show is justified in terms of increasing Deaf visibility on television, disability studies and health communication scholarship on *Switched at Birth* at times glosses over or even entirely ignores race and ethnicity. For example, in an article measuring mass media's role in changing attitudes towards Deaf culture, Seon-Kyoung An et al. write, "Character Bay

Kennish grew up in a wealthy family with two parents and a brother, whereas Daphne Vasquez, who lost her hearing at an early age due to meningitis, grew up with a single mother."[8] The differences between the Vasquez and Kennish families are conceptualized only through number of parents in the household and economic status. In describing the show in this way, the authors foreclose the possibility of considering race and ethnicity intersectionally with their analysis of d/Deaf and hearing characters.

However, paying attention to the racialized representations of both Latinx and white characters on the show is important. Scholars of Latin American studies, communication, and media have all called for increased and better Latinx representation and programming on television.[9] Mary Beltrán notes that there is a long history of non–Latinx actors portraying Latinx characters, or Latinx actors portraying characters from countries of origin other than their own, phenomena which are both seen on *Switched at Birth*.[10] Additionally, she has observed a recent increase in "light-skinned actors of mixed Hispanic heritage" who play Latinx characters but who cannot speak Spanish or have no clear connections to the Latinx community.[11] Although Regina Vasquez (Constance Marie), Bay's biological mother, makes an effort throughout *Switched at Birth* to educate Bay about Latinx culture, especially art, the actress who plays Bay Kennish (Vanessa Marano) is extremely pale-skinned and purportedly of Irish, English, and Italian descent.[12] Such a casting choice is difficult to understand because in the context of the show, Bay's physical appearance is supposed to be something that distinguishes her from the Kennishes and connects her to Regina, yet she reads as exceptionally white to the audience. Certainly, it is possible for a person to identify simultaneously as white and Latinx, but rather than investigating the nuances of such an identity within the show, characters who embody these identities are often positioned as binary opposites. For example, Daphne's academic intelligence contrasts with Bay's artistic passion, playing on racial and ethnic stereotypes that the show suggests are biologically inherited. Furthermore, as seen in the example from the health communication scholarship analyzed above, class status is often conflated with, or stands in for, racial and ethnic positioning. The Kennishes' whiteness at times stands in for their wealth, while Regina's Latinidad stands in for her poverty.

In addition to the previously described qualities, the show also emphasizes the inheritability of phenotypical characteristics. By the time the audience meets them, both Bay and Daphne have long felt different from their families, based on their physical appearances as well as their interests. DNA testing subsequently confirms that this feeling of difference was justified, when each girl learns about the switch. Within today's cultural context, genetic science has a particularly strong institutional authority.[13] Across the

legal, medical, and corporate fields, DNA is often viewed as infallible evidence
that can solve complex social problems. However, as Jay Aronson has shown,
DNA's invincibility is a highly constructed quality, and the ways in which it
has been constructed as such have been obscured. For example, there are
numerous barriers in terms of who is allowed, and in which contexts, to
invoke their right to DNA evidence in criminal cases, but DNA is still imbued
with "mythic infallibility" in determining a person's guilt or innocence.[14] Likewise, there are potential concerns for sample integrity, degradation, and
human and/or laboratory error in DNA testing.[15] However, these issues are
not reported in mainstream media, and thus DNA appears to have a "disem-
bodied power and authority of objective truth."[16] Mass media are one of several social institutions responsible for reproducing reductive versions of
genetic science.[17] When DNA's reified status is not interrogated and when it
is used to address social issues involving marginalized groups, these groups
may become re-marginalized.[18] We see this particularly in medical situations
where so-called "populations" of people are inscribed with particular
diseases.[19] In *Switched at Birth,* for example, this problematically happens
with alcoholism and Latinx characters.

To further analyze these issues, I completed a textual analysis of seasons
one through four of *Switched at Birth* (93 episodes), focusing on representations of race and ethnicity, and the themes of nature versus nurture. Specifically, I considered the contexts in which scientific terms, such as genes,
genetics, biology, DNA, and blood are used to define or to complicate the
characters' identities and their relationships with one another. I also analyzed
the racial and ethnic descriptors Bay and Daphne use for themselves, their
parents, and their communities, as well as those descriptors that others
ascribe to them. Likewise, I noted at which points throughout the show characters' last names are invoked as markers of family, race, and/or class. In
addition to watching the episodes carefully and taking notes on these themes,
I also coded the transcripts for each episode in NVivo to ensure that I did
not miss any instances of the aforementioned terms.[20]

I will begin my analysis by exploring the ways that science is invoked
to explain differences and similarities amongst the family members. Likewise,
I will reflect on the moments when science is undermined, or brushed aside
in favor of cultural explanations for characters' behaviors and traits. Next, I
will consider how racial and ethnic labels, along with last names, are used to
identify characters and their relationships to one another. Finally, I will look
at the ways in which the show first plays into, subsequently disrupts, and
finally reinforces racial and class stereotypes, in a kind of "stereotype bait
and double switch."

Determinations of Science and Culture

The theme of nature versus nurture is central to *Switched at Birth* as an entire series, but is especially prominent in the first season, as the characters come to terms with what has happened. In the pilot episode, "This Is Not a Pipe," Bay Kennish sits in a biology classroom at her expensive private high school, Buckner Hall. Her teacher explains, "From the color of our eyes, to the color of our hair, how your pinky bends, or whether we can roll our tongues—who we are is determined by genes passed down to us from our mothers and fathers."[21] The students then take a blood type test. At dinner that night, Bay shares with her family that she is type AB. Her father, John (D. W. Moffett), explains that this is impossible: both he and his wife, Kathryn (Lea Thompson), are type A, so Bay must be type A or O. He jokes that Kathryn may have had an affair with the mailman. Kathryn responds, feigning insult, that her grandmother was Italian, and "that's where Bay gets her beautiful coloring from."[22] Bay's dark hair and eyes are juxtaposed with her mother and brother Toby's (Lucas Grabeel) red hair and light eyes, though her skin is as pale, if not paler than theirs. "Coloring" here, then, refers to hair and eye color, rather than skin tone. Interestingly, Toby has red hair in the pilot, ostensibly to emphasize that he, Kathryn, and Daphne all share this recessive DNA mutation, but he becomes blonde later in the first season.

Six weeks after this initial conversation, Bay and her parents meet with a genetic counselor who tells them that Bay is not biologically related to her parents, and that there must have been a mix up at the hospital when she was born. John asks what they should do with this new information. Staring straight ahead, Bay answers: "You find your real daughter. I find my real parents." The genetic counselor gently corrects her, explaining, "We say 'biological,' not 'real.'"[23] From this moment on, however, differences between Bay and the rest of her family are typically understood as biological and thus interpreted to mean that she is not "really" part of their family. For example, when Daphne and Regina first come to the Kennish home, John and Kathryn crave additional biological confirmation that Daphne is indeed their daughter. They find that Kathryn, Toby, and Daphne are all allergic to kiwi. Kathryn adds that Bay is not allergic to anything, she is "just picky," thus making her an inconvenient outsider once again.[24]

Scientific evidence also explains the absence of Bay's biological father. In the early episodes of *Switched at Birth*, Regina and her mother, Adrianna (Ivonne Coll), refuse to answer questions about him, and Daphne has no memories of him. During "Pandora's Box," Daphne and her close friend Emmett (Sean Berdy) find a guitar case belonging to Angelo Sorrento (Gilles Marini), who they learn is Regina's former partner and Bay's biological father.[25] Inside the case are candid photographs of Bay throughout her life.

Regina reveals she has known about the switch for thirteen years and that she hired a private investigator to follow Bay as a child. When Daphne was first born, Angelo accused Regina of having an affair because Daphne had blue eyes and was fair in complexion, compared to Regina and him.[26] Angelo secretly took a swab of Daphne's saliva and had her DNA tested. The results came back and his fears were confirmed—Daphne was not his biological child. After Daphne's illness and subsequent hearing loss, Angelo left both of them, despite Regina's insistence that she had not had an affair. The boundaries of family were here again drawn by DNA, and Angelo excluded himself based on the evidence, reflecting the cultural belief that DNA has supreme authority as an objective truth-teller.[27]

Confused, Regina decided to have Daphne's DNA tested herself, and learned that Daphne was not biologically related to her, either. In recounting the story, Regina breaks the biological essentialism that has pervaded the series so far. Kathryn asks, "Why didn't you say anything?" Regina responds, holding back tears, "because she was *my daughter* by then. I was an alcoholic, unemployed single parent. I was broke. I had *nothing*. Nothing but this little baby that everyone thought was mine, who I loved so much."[28] Like Kathryn, Daphne is disappointed in Regina. She yells, "You kept me from my family!" Again, Regina redefines family outside of shared genetic material: "I was your family. The hospital made the mistake, not me. You were three years old. You were *mine* by then."[29] The mother-daughter relationship here is conceptualized by the amount of time that Regina and Daphne had already spent together and the bond that they had formed, when she became aware of the switch.

The question of how to define motherhood lingers throughout the first season. In "Las Dos Fridas," Kathryn's mother, Bonnie (Meredith Baxter), comes to visit.[30] Bonnie and Bay have always had an especially close relationship. However, once Bonnie meets Daphne, Bonnie views Bay in a different light. Bonnie brings presents for each of her (now) three grandchildren. Bay is delighted that she has received a book on Latinx art, until Bonnie presents Daphne with a family heirloom: a cameo she had been saving for Bay. Later, Kathryn asks Bonnie why she gave the cameo to Daphne and not Bay. Bonnie reveals that she wanted to "keep it in the family," adding that "blood is blood. And Daphne is your daughter, I mean, look at those freckles!"[31] Throughout the episode, Daphne's freckles, like her red hair, stand in for her physical similarities to Kathryn, and thus her inherited genetic makeup. When Kathryn insists, "Bay is my daughter, too, mom," Bonnie replies with an argument based in biological racism: "I'm just saying—well, in terms of school. How many tutors did you have to use, just to get her through junior high? And even then she barely made it…. And meanwhile, Daphne, with a single mom, and a handicap, and growing up in a ghetto, she made straight As!

That's your daughter.... Biology tells us what we are."[32] Bonnie implies that the familial cultures in which each girl was raised have not contributed to their talents and personalities, but rather, Bay was predestined to fail and Daphne was predestined to succeed, because of "biology." However, within the context of the episode, "biology" clearly means racial and ethnic background. At the end of the episode, Kathryn confronts her mother. She asks how a "little blood test" could change the way that Bonnie looks at Bay. Bonnie insists, "She's not your DNA, Daphne is. You carried her. Regina carried Bay. She has the connection with Bay that you are never gonna have."[33] Here, Bonnie clearly defines motherhood differently from how Regina defines it in "Pandora's Box," described above.[34] Kathryn responds: "You carried me. We share the same DNA and right now, I couldn't feel any less connected to you."[35] Kathryn's response excludes her mother from her family, despite their biological connection. In these two cases, biological relationships are undermined in order to emphasize the noble sacrifices that both Regina and Kathryn have made for their daughters. Overall, however, after the characters learn about the switch, biological understandings of family are the default. Therefore, these examples suggest that only in the face of extreme circumstances, such as the threat of losing a child or overt discrimination, can the bounds of family be expanded past lines of either biological or racial categories.

Another important theme from the first season that is dealt with in both biological and cultural terms is addiction. In the pilot episode, Kathryn learns that Regina has a criminal record with two DUIs.[36] When Kathryn confronts her with this information, Regina eventually admits that she is a recovering alcoholic, but shares that she has nearly twelve years of sobriety. In an act of rebellion at the end of the pilot episode, Bay attempts to buy alcohol underage with fake identification. Regina, John, and Kathryn arrive at the police station in their first act of co-parenting. The Kennishes are extremely upset and punish Bay by grounding her for a week. In the next episode, "American Gothic," Bay laments her punishment to Regina and is surprised when Regina says she would have been twice as hard on her.[37] Regina reveals that Bay "doesn't have the luxury of [just having] a couple of beers" because Regina herself is an alcoholic.[38] Alcoholism is conceptualized as an inheritable biological problem that also affected Regina's father. Regina tells Bay that they are both "wired that way," implying that it is an inescapable future for Bay.[39]

In the second season, Regina relapses.[40] When she returns home from rehab in "Mother and Child Divided," the idea of predetermined alcoholism again resurfaces. Regina tells John and Kathryn that neurobiologists have discovered that addicts' brains are "genetically wired differently."[41] Of course, research has shown that some people are genetically predisposed to alcoholism.[42] In that sense, it is ethically responsible of the show to portray this. At the same time, research has additionally shown that children who grow up in households with addicts are also more likely than the general population to develop addictions, regardless of genetics.[43] Daphne, however, who has lived with Regina all her life, is not protected nor warned in the same way as Bay, and so again, we see that this particular familial trait is defined as only biological. Such a representation is problematic because the show often conflates the idea of biological family with racial and ethnic heritage. In other words, Daphne's relationship to the Kennishes is claimed and read through her whiteness, and the same holds true for Bay's relationship with Regina and her Latinidad. Bay's artistic skill and hot temper, both Latinx stereotypes, are assumed to be inherited from Regina. By suggesting that she may also inherit Regina's alcoholism, *Switched at Birth* risks inscribing addiction as a Latinx characteristic.

The first season also revolves around another addiction: Toby's gambling. Toby and his friend Wilke (Austin Robert Butler) have become avid poker players. As the plot unfolds, the audience learns that Toby cannot stop gambling and has accumulated thousands of dollars of debt. In "The Persistence of Memory," Toby and Wilke steal and sell a chemistry test from school to pay off the money they owe.[44] Concerned about Toby after hearing rumors circulating at Buckner Hall, Bay enlists Regina's help. Regina confronts Toby and explains the mental processes of addiction to him until he admits that he is addicted to gambling. However, in contrast to Regina's situation, there is no mention that Toby's addiction was unavoidable because of an inherent genetic "wiring" default. No one reveals that any Kennish has ever dealt with a gambling problem. In fact, Toby's addiction resolves quickly when he admits his problem to his father, confesses that he stole the test, and sells his musical equipment to pay his debts. He does not go to counseling or rehab. Toby's problem is presented as the result of poor choices and a culture of risky behavior amongst wealthy teenagers. His addiction is occasionally alluded to throughout the series, but by season three, it has become a joke.[45] No one is concerned that Daphne might start gambling due to her shared DNA because Toby's addiction is not conceptualized as a biological, inheritable problem. In contrast, Regina's addiction is talked about in such biological terms that it becomes inscribed in her. Because of the show's messy relationship between race and genetics, addiction is presented as inheritable between Latinx characters, while white characters are not at risk.

Labeling with Race, Ethnicity and Last Names

Throughout *Switched at Birth* both Bay and Daphne must come to terms with their newfound racial and cultural identities and their "should have been" names. Soon after meeting Regina and Daphne, Bay hides out in her car, incognito, spying on their house in East Riverside. A young man, Ty Mendoza (David Blair Redford), approaches her and tells her to leave, saying he doesn't like "rich white girls slumming it looking for dime bags." Bay tells him that she's not looking for pot and that she's not a "rich white girl."[46] Here is the first time that Bay denies her whiteness, despite her extremely pale skin. Of course "whiteness" as an ideological construct means much more than the pigment of one's skin, and throughout the series, both Bay and Daphne will grapple with this idea. However, I draw attention to Bay's pallor because it is impossible to believe, within the context of the show, that she would not benefit from white privilege, either before or after the switch comes to light.

Additionally, the show never fully interrogates the idea that whiteness and Latinidad can be co-inhabited identities, but rather sets them up in a binary opposition. In their work on colorism in Latina communities, Laura Quiros and Beverly Araujo Dawson found that their participants understood racial and ethnic identification in myriad ways. Amongst women with lighter skin, some identified as Latina and white, while others recognized the benefits that light skin conferred on them, but did not claim whiteness as a racial identity. Some women explicitly conceptualized race in terms of a black/white binary and viewed their Latinidad as outside of that binary, as an ethnicity. Overall, whiteness and Latinidad were not dominantly understood as mutually exclusive categories, as they are typically portrayed in *Switched at Birth*.[47]

Jorge J. E. Gracia suggests that today's increasingly diverse society presents both factual and epistemic challenges to the legitimacy of the categories of race and ethnicity. Latinx identities in particular show that these groupings are not stable across time nor geography, and that many people cannot be easily classified through either system.[48] American society is still working these ideas out, as we can see in situations where people explicitly need to choose labels, such as on censuses and job applications. In *Switched at Birth*, Daphne and Bay are often positioned as opposites, with Daphne read as white and Bay as Latina. Due to their unique situation, both Daphne and Bay could also potentially be read as disrupting these categories, though the show only rarely explores this idea.

Bay's early attempts to understand her "new identity" result in her enactment of ethnic and class stereotypes. The morning after Regina and Daphne come over for dinner, Bay sits at the breakfast counter.[49] She has pierced her nose, is drinking coffee, and pulls out a cigarette. When Kathryn questions this behavior, she explains that she is "just living the life I was supposed to live." She interprets her new Latina identity as fun and wild, culminating in her arrest at the liquor store, where she tells police that her name is Daphne Vasquez. Bay continues to understand her Latina identity through reckless behavior for several episodes. However, when her Kennish grandmother comes to visit in "Las Dos Fridas," for the first time, Bay understands her ascribed racial difference negatively. Holding back tears, she explains to Kathryn: "I'm not like her. I'm not like you. I'm a whole other race. My name is supposed to be Vasquez and Grandma is not the last person that's going to think that that automatically means that I'm bad at school, or illegal, or whatever. Mom, I have never had to think about what that feels like, until now and—have you?"[50] Here, Bay confronts her white privilege, in a sense. She realizes she has been fortunate not to have been discriminated against based on her last name. However, she fails to acknowledge that her last name still is Kennish, and that her physical appearance does not change to the outside world just because she has learned new information about her ancestry. In this instance, *Switched at Birth* misses an opportunity for explicit interrogation of whiteness in the context of a "nature versus nurture" debate. This scene firmly places Bay back into the role of "rich white girl," albeit one who is beginning to see the world from a broader perspective.

Like Bay, Daphne undergoes an identity shift during "Las Dos Fridas." By the middle of the first season, Daphne plays basketball for Buckner Hall and has tentatively befriended Bay's ex-best friend, Simone (Maiara Walsh).[51] While in the locker room, Daphne sees an old friend she knows from "the barrio," Monica (Natalie Amenula). Monica explains that the coach at Buckner Hall lets the track team from Daphne's old neighborhood use the good quality track at Buckner because he suffers from "white guilt."[52] Daphne and Monica laugh. As they are reminiscing, Simone enters the locker room and accuses Monica of stealing her watch, implying that Monica would do so because she is poor and could pawn it for money. Monica denies the accusation, and she and Simone end up physically fighting. As punishment for the altercation, the basketball coach assigns Simone and Daphne community service in East Riverside so that they can learn to "respect the people that live there."[53]

While Simone and Daphne are completing their assignment (painting over graffiti, which itself may be viewed as a colonizing practice), Monica approaches them, confrontationally. She tells Daphne not to act superior, just because she lives in Mission Hills now. Daphne tries to talk her down, saying, "I'm the same Daphne I was before. My name's still Vasquez." Monica replies, "'Til you change it to something white. You never looked like any Vasquez I ever knew, anyway."[54] Daphne is visibly shaken by this response, and the audience sees that she, like Bay, must make sense of her new social position. Daphne's realization that the way she views herself is not the way that the rest of the world views her is perhaps more earnest than Bay's. It is probably true that most of the world would not assume that Daphne is Latina, apart from her last name. Daphne, like Bay, must reconcile her own whiteness.

Daphne grapples with this theme of her "internal" identity not matching her phenotypical appearance throughout the series. By season three, she plans to attend Gallaudet University, a university for d/Deaf and hard of hearing students, which is "affordable" in the grand scheme of U.S. college tuitions. However, in "Oh, Future!" Daphne tells Regina that she wants to expand her list and apply to some expensive colleges to study pre-med.[55] Regina has put some money aside, but shares that it will not be enough for even one semester at an elite private school. Daphne looks for scholarship programs, but they are difficult to find. Later, John and Kathryn sit down with Regina and offer to pay for Daphne's education. Regina is insulted and insists she doesn't want Daphne to have everything handed to her on a silver platter. At the same time, she feels guilty that she cannot afford the expensive schools Daphne wants to attend, so she does her own research and discovers that there is a Kansas Latina Merit Award worth $25,000. Despite Daphne's protest that she cannot apply, Regina insists that she is eligible, saying, "You are Latina because that is the culture you were raised in."[56] Here, Regina defines Latinidad as a cultural, rather than racial or ethnic, category. Later in the episode, Daphne tells her friend, Sharee (Bianca Bethune), who is Black, about her plans to interview for the scholarship. Sharee responds incredulously, "Excuse me, you're not Latina." Daphne refutes Sharee's argument by claiming, "My last name is Vasquez." Sharee attempts to explain Daphne's white privilege to her: "If you walk into a store with a big coat on and some girl with dark hair and dark skin walks into a store with a big coat on, you're not gonna have ten salespeople follow you around. She will."[57] Daphne insists, "But that's about skin color, not about being Latina. Some Latinas are blonde, some are redheads."[58] Here again was an opportunity for the show to push harder at the idea that a person may benefit from white skin privilege and still be Latina. However, the conversation stops short. Despite feeling conflicted, Daphne attends the interview. During the conversation, she discusses the hardships she has encountered from being Deaf. The interviewer empathizes and asks, "Can you think of a time where you felt similar discrimination because you're Latina?" When Daphne cannot think of an example, the woman kindly explains, "This scholarship was started to help girls who had been overlooked or passed over because they're Latina." Daphne angrily retorts, "So I'm not Latina if I haven't been a victim because of it?" The interviewer responds, "I'm not saying that. We're not celebrating victimhood." Daphne insists, "Being Puerto Rican is a big part of who I am," yet relents, and realizes that this scholarship was not meant for her.[59]

When Daphne returns home, she confesses to Regina that she withdrew her application because she realized that she should not have applied in the first place. Daphne, finally recognizing her social position, shares, "If I won, I wouldn't feel right about taking it from someone who the whole world saw as Latina." Regina angrily counters, "So they deserve it more because they have dark hair and brown skin? So you're not Latina because you don't feel burdened enough? Me and your grandmother and generations before us faced discrimination because our name is Vasquez. You don't think that affects you?" Daphne responds quietly, "Yeah, my name is Vasquez, but by blood, I'm a Kennish, whatever they are."[60] Daphne's appeal to her "blood" again emphasizes that "real" racial and ethnic identity is biological. Although Daphne's insistence on not applying for the scholarship may seem moral, in the next season she uses her claimed cultural Latinidad to gain entry into a college class that is full, as we saw in the opening vignette of this essay. To its credit, *Switched at Birth* is not afraid to present the characters with ethical dilemmas such as this one. At the same time, neither Bay nor Daphne ever show a full understanding of their privilege, and their invocations of Latinidad as hardship when convenient come across as insensitive and insincere.

Breaking and Reinforcing Stereotypes

Admirably, *Switched at Birth* presents each of its characters in multifaceted ways. None of the main nor recurring characters are depicted as entirely good, or entirely evil. However, there is a pattern in the way some of the women of color are represented that undermines this overall complex character development. Both Regina and Hope (Kim Hawthorne), the mother of the son of a man Regina dates in the fourth season, are subjected to what I will call a "stereotype bait and double switch." In this narrative move, the character is first presented as consistent with racial and ethnic stereotypes, but soon audience members learn that she is actually a noble person and thus feel guilty about making racist assumptions. Once this is resolved, however, the show reverts to these negative stereotypes, and audience members are left with the feeling that their racist preconceptions were right all along.

Regina is subjected to this move early in the first season. She is set up as a stereotypical poor Latina woman: she is a single mother who lives in a dangerous neighborhood with "bars on the windows. Bail bonds on every corner."[61] She drives an old, beat up car. She could be described as having a temper. However, once the audience learns that Daphne is an excellent student, talented athlete, and kind person, Regina's role shifts from stereotypical poor Latina woman to heroic mother of a child with a disability. She is a "good" or "model" minority. Such a narrative trope is common in a society that considers itself "post-racial" or "colorblind," but which does nothing to actually improve life for marginalized groups.[62] Yet the show does not stop there. Soon, as previously described, Kathryn discovers that Regina has two

DUIs. Next, John and Kathryn find out that Regina has known about the switch for over a decade and never contacted them. They assume she wants financial compensation. Within the first several episodes, therefore, Regina is labeled with the negative, stereotypical traits of addiction, criminality, and untrustworthiness.

In season four, Regina becomes co-owner of a coffee shop. She begins dating, then subsequently moves in with her business partner, Eric (Terrell Tilford). Eric, who purports to be a widower, has a young son named Will (Sayeed Shahidi) who also lives with them. Regina is aware that something is not quite right about Eric and his mysterious past, but decides to overlook it. However, when Will confides in Regina that his mother is still alive, Regina confronts Eric. In "To Repel Ghosts," Eric admits that he kidnapped Will from his ex-wife because she was an abusive drug addict who neglected her child. He and his son are now living under fake identities in Kansas City.[63] Here, we see the first move of the "stereotype bait and double switch," wherein Hope is depicted as a stereotypical unfit, drug-addicted Black mother. Regina wants to believe Eric's story, but plans a secret trip to Atlanta, where Hope currently resides, to investigate for herself. Regina and Daphne fly to Georgia, where Regina rents a car. She tampers with the water pump hose so that she has an excuse to bring the car to the auto mechanic shop where Hope works. While Hope repairs the hose, she and Regina begin to talk. They bond over being parents and Regina subtly hints that she is an addict. Hope informs her that she is a member of Narcotics Anonymous, and invites her to a meeting. Unsurprisingly, the stereotype is debunked, and Daphne and Regina now believe that "Eric was totally wrong about her."[64]

However, after the meeting, it is revealed that Hope and her current husband have a scheme wherein they pretend to be recovering addicts and gain the trust of members in the NA group, but then sell drugs to these vulnerable people in the parking lot after meetings. Again, as in Regina's case, Hope is now reinscribed as an addict, a liar, and a criminal. While such narrative moves prevent the show from relying too heavily on post-racial tropes of individual "good minorities," they are also somewhat socially irresponsible. As previously discussed, the theme of addiction is developed throughout *Switched at Birth*, alongside the theme of genetic predisposition to certain behaviors. While each of these themes is interesting to consider within the show's context, their simultaneous presentation and articulation with racial stereotypes leads to the problematic representations described here.

In considering both of these cases, Beltrán's concept of meaningful diversity is a helpful analytic tool. After all, each woman has fully developed, complex back stories. In fact, Beltrán warns against being too afraid of reproducing negative stereotypes to the extent that any sense of racial difference is lost. She suggests that this fear is partially responsible for the recent influx in light-skinned Latinx representations in which characters are Latinx by name only, and not any cultural or linguistic association.[65] Therefore, the "stereotype bait and double switch" is not necessarily always a "bad" narrative device. However, it is important to acknowledge here that Regina and Hope, both women of color, are the two characters on which this move is developed most clearly. Given that the show sometimes fails to distinguish between racial and biological categories, this is problematic and risks suggesting that these stereotypes are inherent qualities.

Conclusion

As we have seen, *Switched at Birth* deals with the categories of race and ethnicity in ways that both support and undermine the nature versus nurture binary, and which rupture, yet also reinforce, racial and ethnic stereotypes. The pervading theme of "biological as real" and the insistence on DNA as an arbiter of truth means that although one goal of the show may be to redefine what family means or can look like, biological family and identities are ultimately viewed as more legitimate than chosen ones.

Switched at Birth succeeds in presenting an overall meaningfully diverse cast, yet it also contains some problematic representations. While the characters sometimes view their traits as cultural products, such as when Kathryn stands up to her racist mother and insists Bay is her daughter, negative traits, such as addiction, are viewed as unavoidably biologically transmitted, and only amongst characters who are people of color. Furthermore, both Daphne and Bay invoke their Latinidad when it is convenient, and never fully interrogate their whiteness or white privilege. In fact, each often tries to distance herself from it. Finally, with the "stereotype bait and double switch," two actresses of color on the show are subjected to an especially insidious kind of representation, which may reaffirm audience members' racist stereotypes.

Switched at Birth has been rightly praised for greatly increasing the visibility of Deaf youth culture in mainstream media. Additionally, it takes admirable risks in dealing with the issues of race, ethnicity, and genetics, rather than just ignoring them and continuing to propagate a colorblind, post-racial ideology. It also opens up interesting conversations regarding the relationships between DNA and our social identities, in our increasingly technologically advanced society. Nevertheless, the show at times dangerously conflates genetics with race and ethnicity and makes use of racial stereotypes to prove that biological relationships are "real." Moving forward, it is important to continue to theorize how television producers and writers can create characters with intersectional identities in ways that highlight their complexities and engage in current social issues, without falling back on reductive stereotypes.

NOTES

1. "And It Cannot Be Changed," *Switched at Birth*, season 4, episode 1, ABC Family, January 6, 2015.

2. See, for example, Kathryn A. Foss, "(De)stigmatizing the Silent Epidemic: Representations of Hearing Loss in Entertainment Television," *Health Communication* 29 (2014): 897–898; Elizabeth Ellcessor, "'One Tweet to Make So Much Noise': Connected Celebrity Activism in the Case of Marlee Matlin," *New Media & Society* (2016): 1–17.

3. Kathryn A. Foss, "Constructing Hearing Loss or 'Deaf Gain?' Voice, Agency, and Identity in Television's Representations of d/Deafness," *Critical Studies in Media Communication* 31, no. 5 (2014): 440.

4. Deaf activists and bloggers, along with ASL teachers, have critiqued specific aspects of the show, such as their misrepresentation, and lack of representation of cochlear implants ("Switched at Birth's First Big Mistake: An Inaccurate Representation of Cochlear Implants," *Redefined Magazine*, July 9, 2013, http://www.redefined.com/2013/07/switched-at-births-first-big-mistake.html), and the level of technique with which the hearing actors sign (David Boles, "Total Failure of the ASL-Only 'Switched at Birth' Episode on ABC Family," March 5, 2013, https://bolesblogs.com/2013/03/05/total-failure-of-the-asl-only-switched-at-birth-episode-on-abc-family).

5. Melanie E. S. Kohnen, "Cultural Diversity as Brand Management in Cable Television," *Media Industries Journal* 2, no. 2 (2015): 88; Mary Beltrán, "Meaningful Diversity: Exploring Questions of Equitable Representations on Diverse Ensemble Cast Shows," *Flow* 12, no. 7 (2010).

6. Kohnen, "Cultural Diversity as Brand Management in Cable Television," 88.

7. *Ibid.*, 89.

8. Seon-Kyoung An, Llewyn Elise Paine, Jamie Nichole McNiel, Amy Rask, Jourdan Taylor Holder, and Duane Varan, "Prominent Messages in Television Drama Switched at Birth Promote Attitude Change Toward Deafness," *Mass Communication and Society* 17 (2014): 205.

9. Mary Beltrán, "Latina/os on TV! A Proud (and Ongoing) Struggle Over Representation and Authorship," in *The Routledge Companion to Latina/o Popular Culture*, ed. Frederick Luis Aldama (Routledge: New York, 2016), 23–33; Mari Castañeda, "The Importance of Spanish Language and Latino Media," in *Latina/o Communication Studies Today*, ed. Angharad N. Valdivia (New York: Peter Lang, 2008), 51–66.

10. Beltrán, "Latina/os on TV!" 26–30.

11. *Ibid.*, 31.

12. Marano's ethnicity is purported to be Irish, English, and Italian on sites such as http://ethnicelebs.com/vanessa-marano.

13. Brian Wynne, "Reflecting Complexity: Post-Genomic Knowledge and Reductionist Returns in Public Science," *Theory, Culture and Society* 22, no. 5 (2005): 70.

14. Jay D. Aronson, "Certainty vs. Finality: Constitutional Rights to Post-Conviction DNA Testing," in *Reframing Rights: Bioconstitutionalism in the Genetic Age*, ed. Sheila Jasanoff (Cambridge: MIT Press, 2011), 134.

15. Aronson, "Certainty vs. Finality," 135.

16. *Ibid.*, 127–128.

17. Wynne, "Reflecting Complexity," 75–76.

18. Jenny Reardon, "Human Population Genomics and the Dilemma of Difference," in *Reframing Rights: Bioconstitutionalism in the Genetic Age*, ed. Sheila Jasanoff (Cambridge: MIT Press, 2011), 221.

19. See, for example, the concept of "bioethnic conscription" in Michael J. Montoya, *Making the Mexican Diabetic: Race, Science, and the Genetics of Inequality* (Berkeley: University of California Press, 2011), 185.

154 A New Kind of Family

20. Transcripts were accessed via http://transcripts.foreverdreaming.org/viewforum. php?f=158.

21. "This Is Not a Pipe," *Switched at Birth*, season 1, episode 1, ABC Family, June 6, 2011.

22. *Ibid.*

23. *Ibid.*

24. *Ibid.*

25. "Pandora's Box," *Switched at Birth*, season 1, episode 8, ABC Family, July 25, 2011.

26. Angelo describes his heritage as "Italian, French, some Algerian," to which Bay replies, "Woah, like I'm Arab?" during "The Homecoming." *Switched at Birth*, season 1, episode 10, ABC Family, August 8, 2011.

27. Aronson, "Certainty vs. Finality," 127–128.

28. "Pandora's Box," *Switched at Birth*.

29. *Ibid.*

30. "Las Dos Fridas," *Switched at Birth,* season 1, episode 16, ABC Family, February 7, 2012.

31. *Ibid.*

32. *Ibid.*

33. *Ibid.*

34. "Pandora's Box," *Switched at Birth*.

35. "Las Dos Fridas," *Switched at Birth*.

36. "This Is Not a Pipe," *Switched at Birth*.

37. "American Gothic," *Switched at Birth*, season 1, episode 2, ABC Family, June 13, 2011.

38. *Ibid.*

39. *Ibid.*

40. "Drive in the Knife," *Switched at Birth*, season 2, episode 7, ABC Family, February 18, 2013.

41. "Mother and Child Divided," *Switched at Birth*, season 2, episode 11, ABC Family, June 10, 2013.

42. Arpana Agrawal and Michael T. Lynskey, "Are There Genetic Influences on Addiction: Evidence from Family, Adoption and Twin Studies," *Addiction* 103, no. 7 (2008): 1072.

43. National Institute on Alcohol Abuse and Alcoholism, "A Family History of Alcoholism: Are You at Risk?" 2012, https://pubs.niaaa.nih.gov/publications/familyhistory/famhist.htm.

44. "The Persistence of Memory," *Switched at Birth,* season 1, episode 6, ABC Family, July 11, 2011.

45. "And We Bring the Light," *Switched at Birth*, season 3, episode 15, ABC Family, June 30, 2014. The analysis in this essay covers seasons 1–4 in *Switched at Birth*. However, in season 5, Toby's gambling problem resurfaces. It is blamed on his living in the UK, where gambling is seen as a more socially acceptable activity.

46. "This Is Not a Pipe," *Switched at Birth*.

47. Laura Quiros and Beverly Araujo Dawson, "The Color Paradigm: The Impact of Colorism on the Racial Identity and Identification of Latinas," *Journal of Human Behavior in the Social Environment* 23 (2013): 293.

48. Jorge J. E. Gracia, "Race or Ethnicity? An Introduction," in *Race or Ethnicity?: On Black and Latino Identities*, ed. Jorge J. E. Gracia (Ithaca: Cornell University Press, 2007), 1–4.

49. "This Is Not a Pipe," *Switched at Birth*.

50. "Las Dos Fridas," *Switched at Birth*.

51. Though the character of Simone is likely read as white and non-Latinx by the audience, due to her wealth and light skin, the actress who portrays her is purportedly of both Western European and Brazilian descent: http://ethnicelebs.com/maiara-walsh.

52. "Las Dos Fridas," *Switched at Birth*.

53. *Ibid.*

54. *Ibid.*

55. "Oh, Future!" *Switched at Birth,* season 3, episode 14, ABC Family, June 30, 2014.

56. *Ibid.*

57. *Ibid.*

58. *Ibid.*

59. *Ibid.*

60. *Ibid.*

61. "This Is Not a Pipe," *Switched at Birth.*

62. Riva Tukachinsky, Dana Mastro, and Moran Yarchi, "Documenting Portrayals of Race/ Ethnicity on Primetime Television over a 20-Year Span and Their Association with National-Level Racial / Ethnic," *Journal of Social Issues* 71, no. 1 (2015): 33.

63. "To Repel Ghosts," *Switched at Birth*, season 4, episode 11, ABC Family, August 24, 2015.

64. *Ibid.*

65. Beltrán, "Latina/os on TV!" 32.

A Different Kind of Foster Family

Portrayals of Teen Foster Care on Freeform

PATRICE A. OPPLIGER *and* MEL MEDEIROS

The Freeform network has an intriguing history, from Christian Broadcasting Network to family-friendly The Walt Disney Company–owned ABC Family.[1] In 2006, the channel adopted the slogan "A New Kind of Family" and began "inching toward edginess."[2] In 2015, renamed "Freeform," the channel focused its programming on "the Becomer audience," i.e., viewers from 14 to 34 years old.[3] ABC Family president Tom Ascheim announced that "Freeform will deliver new, exciting original content as well as all the favorite shows our viewers already love on ABC Family."[4] Conservative critics, on the other hand, see the name change as positive since viewers, "once hoodwinked by the 'Family' moniker," would be not able to easily avoid the "edgy" content.[5]

Upon announcement of *The Fosters* in 2013, a show which features a biracial lesbian couple raising foster, adopted, and biological children, conservative group One Million Moms propagated an email-writing campaign to its sponsors, protesting the show's LGBT content. The sample email suggested viewers voice their objections such as "As a Christian, I feel you are attacking my faith, morals and values."[6] The organization called for families to avoid Freeform entirely; otherwise, they ran the risk of viewing promos for this new, "offensive" program.[7] One Million Moms was concerned that *The Fosters* would desensitize audiences to liberal views and promote "inappropriate" behavior such as same-sex coupling.[8] They believed the show was "degrading to families and damaging to our culture."[9] The group further argued that the show's positive emphasis on helping foster children was offset by attempts to redefine marriage and family. However, the negative publicity was apparently not damaging to ratings—the premiere of *The Fosters* attracted 1.9 million people and was cable's top show among its target audience of females ages 12 to 34. Executive Producer Peter Paige stated, "A stamp of disapproval from One Million Moms is like a critic's pick."[10]

The Fosters creators Bradley Bredeweg and Peter Paige reported ABC Family was receptive to the idea of a show about lesbian, multi-racial parents from the very beginning. The network's slogan "A New Kind of Family" fit the model of the show.[11] We argue that the "new kind of family" approach is moral, in contrast to conservative religious groups' claims of impropriety. It is important for young viewers, especially those from nontraditional families, to see their lives reflected in media portrayals. According to a Pew Research Center report, 46 percent of U.S. children under age 18 are living in a home with two married heterosexual parents in their first marriage.[12] Since a "normal" family is itself not a majority, no child should see their family, whether led by single, same-sex, or foster parents, as abnormal or damaged. The Freeform shows are powerful for their ability to positively and accurately reflect nontraditional families and show both the struggles and successes of the teens.

Freeform programming addresses social issues not commonly found on adult-targeted channels. One such issue is the foster care system where, according to government reports, "on any given day, there are approximately 415,000 children in foster care in the United States. In 2014, over 650,000 children spent time in U.S. foster care."[13] Between 2013 and 2016, the number of children in foster care rose 3.5 percent, likely because of the increase in cases of parental opioid addiction.[14] The statistics on foster children are often chilling, especially for the 26,000 young adults who age out, reaching the age at which the government decides they are legally adults and no longer can receive support through foster care, of the system each year.[15] For those who age out without a support system, one in five will become homeless, one in four will be incarcerated within two years, and, for young women, 71 percent will become pregnant before they reach the age of 21.[16]

We argue that the four Freeform shows in this analysis accurately and realistically portray foster care from the perspectives of those experiencing the system firsthand. Adults, from foster parents to social workers, whether caring or malicious, are often presented with a cool objectivity, representing the rather sterile bureaucracy of the foster care system. The teens in the foster system, on the other hand, are portrayed with sensitivity and nuance: the teens are neither entirely good nor inherently bad. Instead, these Freeform programs portray foster teens' lives, showing the justifications and provocations for their actions, as much as the consequences. The character-driven storylines engender empathy and understanding for youth in foster care without simplifying their stories to ones only of victimhood. In this essay, we will analyze the two sides of these portrayals, the objective narratives of the foster care "system" and the nuanced, genuine stories of the kids who experience it.

Social Representation Theory

The theory of social representations illuminates the impact of programming like *The Fosters* on audiences, explaining that individuals gathering information about an unknown phenomenon from the mass media is the first step in forming public opinion about that phenomenon.[17] Representations are not defined by any particular individual such as a film director, television producer, or media critic, but rather, are collectively and socially formed by groups through interpersonal communication and media portrayals.[18] Viewers combine the new information they learn through media with familiar experiences encountered in their daily lives. Representations are also dynamic and constantly being created and recreated by media and society.[19] Not only do attitudes shift as media portrayals push boundaries of traditional values (e.g., beyond heteronormative, cisgender characters), television shows and movies spark discussion and debate among fans as divergent portrayals are introduced. In turn, new meanings are created, and the abstract becomes familiar.

Social representations are particularly important for prevalent but indirectly experienced issues, such as foster care, which affects many children and families, yet few outside the system have personal exposure to it. Foster children are often moved from home to home, not staying long enough in schools or neighborhoods to make connections. The children may also hide their status out of shame or fear of discrimination. Often, what people know about the foster system, they learned from news reports or fictional portrayals in the entertainment field.[20] Because foster care is often invisible to outsiders, it is important to explore its media portrayals. Images of foster children are infrequent in entertainment media, with realistic depictions even scarcer. Examples of foster children who appeared on network television include Samantha (Ashley Rickards) on *One Tree Hill* (WB/The CW, 2003–2012), Lux (Britt Robertson) on *Life Unexpected* (The CW, 2010–2011), Rickie (Wilson Cruz) on *My So-Called Life* (ABC, 1994–1995), Eric (Blake Bashoff) on *Judging Amy* (CBS, 1999–2005), and Santiago (Benjamin Ciaramello) on *Friday Night Lights* (NBC/101 Network, 2006–2011). More often, media depict foster children as entirely damaged.[21] In extreme cases, such as Natalie (Jessica Collins) from season 7 of *CSI: Crime Scene Investigation* (CBS, 2000–2015), Gary (William Lee Scott) and Erwin (Francis Capra) in *Criminal Minds* (CBS, 2005–present),[22] and Chris (Wil Horneff) in *Law & Order* (NBC, 1990–

2010),[23] former foster children are so broken they became killers. By presenting the stories of foster children solely through the lens of the difficulty they have faced, success is presented as an unlikely outcome for those children.[24] Social representations are essential to communication, but negative representations can have adverse impacts on minority or otherwise marginalized groups.

Since people not directly involved in foster care have mostly mediated and indirect experiences with the child welfare system, media portrayals of foster care are likely to have a substantial influence on the public's perception of the issue and its willingness to get involved. Most Americans rely on sporadic news coverage of the issue, which is often limited to accidents and wrongdoing.[25] Significant negative coverage of the foster care system has been found in an analysis of Australian news, a system which is similar to the United States.[26] Negative portrayals leading to and reinforcing negative stereotypes may have significant consequences, contributing to widespread images of foster care as an inherently negative experience for all parties involved.[27] Further, these adverse perceptions of foster care have been associated with problems recruiting and maintaining foster families,[28] which in turn makes the absence or negativity of foster care portrayals significant.[29] Despite a nationwide decline in the number of children in foster care between 2008 and 2013, states still faced foster parent shortages as of 2016. At the same time, veteran foster families retire, and it is becoming more difficult to recruit new families.[30] The shortage—especially of families willing to take in teens— has devastating consequences for society, resulting in higher pregnancy rates, unemployment, and drug use, as well as limited access to higher education and the opportunities it provides for former foster children.[31] Since foster care in the U.S. is publicly funded, negative public opinion can affect legislation and funding, as well as foster parent recruitment.

The Entertainment Industries Council (EIC) produced a special publication illuminating the role media can play in advancing positive foster care.[32] The EIC hosted the "Picture This: Foster Care" panel, a collaboration with various foster care experts from the Casey Family Foundation. They met to establish "a consensus on depiction priorities and recommendations for entertainment and news media to address foster care."[33] The EIC further claims that popular entertainment has "undeniable power" to generate awareness and motivate viewers to pay attention, shape perceptions, and affect emotions. Skillful storytelling with identifiable characters that taps into viewers' emotions is important in influencing attitudes.[34] Entertainment married with education provides a powerful means to present role models, especially within the foster care system. Such portrayals can motivate viewers to become involved and advocate for policy and program change.

A former foster child and current television reviewer, Chris Chmielewski, admitted to cringing when learning about fictional foster care portrayals because foster care is often used for sitcom fodder or as a *Law & Order* backdrop, and Hollywood "rarely gets it right. It's either too overdramatized like *Annie*, too intense like *Shameless*, or maybe too rosy and fairy tale-ish like *Diff'rent Strokes*. They are fine for what they are, but that isn't reality."[35] Because of intense scrutiny of the foster care system, television shows such as *The Fosters* strive for accurate and authentic portrayals of the system. *The Fosters'* producers consulted with organizations such as the Foster Care Alumni of America (FCAA). Indeed, the non-stereotypical portrayals of and storylines with foster children on Freeform work to shatter misconceptions

Analysis

ABC Family/Freeform programs *Pretty Little Liars* (2010–2017), *The Lying Game* (2011–2013), *The Secret Life of the American Teenager* (2008–2013), and *The Fosters* (2013–2018) have featured major characters in the foster system. This essay explores portrayals of foster care in dramas targeted to teenagers and young adults. This age group is important since nearly 50 percent of the 415,000 children in foster care are over the age of 10.[36] Programs on Freeform—*The Fosters* in particular—provide a nuanced view of teenagers in foster care, who are often presumed to be difficult to handle and too old to change.[37] In this analysis, we address foster care as a plot device, the tropes and stereotypes associated with the foster system, and its potential influences on public opinion.

Even within the Freeform network, portrayal of foster care is not uniform; each series portrays foster characters in distinct ways. In *The Secret Live of the American Teenager*, Ricky (Daren Kagasoff), the son of an alcoholic mother and a sexually abusive father, has lived in several foster homes before ending up in the stable home of Margaret (L. Scott Caldwell), an African American social worker, and Sanjay (Brian George), an Indian American physician. At 16, Ricky impregnates a 15-year-old classmate. *Pretty Little Liars* (2010–2017) includes a secondary character, Caleb (Tyler Blackburn), whose biological mother abandoned him when he was five and who has run away from one of the foster homes where he was placed. Drama arises when Caleb becomes romantically involved with one of the main characters, Hanna (Ashley Benson). In *The Lying Game*, identical twins Emma and Sutton (both played by Alexandra Chando) are separated at birth and raised in two different families: one in a stable adopted family, one moved from foster family to foster family. Emma has been placed in six homes and attended a dozen schools between ages 3 and 16. In her latest placement, Emma's foster brother Travis (Kenneth Miller) stole money and blamed Emma.[38] Clarice (Debrianna

Mansini), Emma's foster mother, sides with her biological son and calls the police, forcing Emma to flee the home.

The Fosters, although it primarily features 16-year-old Callie (Maia Mitchell), differs from the other Freeform shows in that it includes stories of multiple foster children. About two-thirds of children in foster care have siblings who are also in the foster system.[39] Callie and her younger brother Jude (Hayden Byerly) had been in multiple foster homes, many of which were abusive, in their journey to a forever family, the Adams-Fosters. Jude and Callie entered foster care after their mother was killed in a car accident caused by their drunk father, who was sent to prison. Callie's foster siblings in the Adams-Foster household, Mariana (Cierra Ramirez) and Jesus (Jake T. Austin, 2013–2015; Noah Centineo, 2015–2017), are former foster children themselves who came into the system when their drug-addicted birth mother abandoned them. Unlike the Freeform shows discussed earlier, which used foster care as isolated plot points, *The Fosters* uniquely focuses on it, showing the foster system through several characters who each entered and experienced foster care differently in loving foster families, abusive foster homes, group homes, and juvenile detention. The main character Callie is presented not as "a lurid case study," but rather is portrayed realistically and with sensitivity.[40]

The research and diligence that went into creating the foster care portrayals on the Freeform network is evident, especially in *The Fosters*. Creators Bredeweg and Paige aimed to create an educational and entertaining show that "is not only opening hearts and minds to equality and social issues, but it is also helping fans imagine the good they might do for foster youth who may be in their tweens or teens."[41] Bredeweg's inspiration came from a childhood friend who struggled in the foster system until she was eventually adopted.[42] Bredeweg stated that he and Paige realized they had a story that had not been told before on television.[43] Paige explained that immersive foster care stories were previously untold because "it's hard for us as a culture to look at the ways that we're failing…. We're failing some kids. There are kids without homes. How is that okay?"[44] Bredeweg and Paige expressed their concern with accurately representing the foster care system, taking great care to accurately portray social workers, foster parents, birth parents, and foster teens.[45]

The System

The Freeform shows portray the vulnerability that children experience in the foster care system, where they rely on adults to keep them safe. The portrayals are realistic in that they depict common situations, such as social workers scrambling for a last-minute placement that might only last for a night or two before displaced children are shuffled off to yet another home. A focus group of former foster children noted they appreciated *The Fosters'* sensitive depictions of coping with abuse and placements in unfamiliar homes with strangers.[46] For example, when Callie arrives at yet another foster home, she interrupts what she sees as a litany of rules and rudely informs her new foster moms Stef (Teri Polo) and Lena (Sherri Saum) that this is not her first foster home. Presuming she will not be there long, Callie simply wants to know where she will sleep that night. When Lena asks Callie if she has a toothbrush, Callie curtly responds, "No, I don't have a toothbrush. How would I have gotten a toothbrush?"[47] It appears Callie left the juvenile facility with nothing but the clothes on her back. It is common for foster children to be taken out of their homes without notice: the few things they are allowed to take are hurriedly placed in garbage bags.

Mirroring real life politics, *The Fosters* addresses the highly contentious issue of private, for-profit foster homes. The character Justina Marks (Kelli Williams) is a lobbyist who promotes a foster care reform bill that will privatize much of the housing. To illustrate how for-profit organizations often fail to be diligent about checking the foster home's history, a 14-year-old boy named Jack (Tanner Buchanan) is moved from an abusive group home to one of these privatized homes where he ends up dying at the hands of his abusive foster father,[48] who happens to be the same man who abused Jude. Once Callie realizes the harm the bill could do, she and the Adams-Foster family work to defeat Justina's bill and replace it with a bill called Jack's Law.[49]

The Fosters also delves into the dynamics of group homes. While the show touches briefly on a boys' group home where Jack is bullied, significant screen time is given to the girls-only group home Girls United. The range of personalities and circumstances of its residents allows *The Fosters* to tell more stories of foster care. For example, Becka (Annamarie Kenoyer) was placed in foster care after having been sexually abused by her mother's boyfriends and turning to drugs—with the aid of her grandmother—at age eleven. Cole (Tom Phelan) was sent to the group home for stealing and prostitution after his parents kicked him out for being transgender. However, not all of the Girls United residents are in the child welfare system because of neglect or abuse by their parent. Other characters committed various crimes, and the line between the foster care system and the juvenile justice system is blurred in the show as in real life. Residents were placed in Girls United because of vandalism, selling drugs, gun possession, gang activity, cyberbullying, home invasion, and robbery.

Girls United also presents a diverse population. Although television series often fail to represent minority characters, *The Fosters* reflects many real life demographic trends. Approximately 41 percent of children in foster care are White, 27 percent are Black, 21 percent are Hispanic, and the remaining 8 percent are multiracial (5 percent), American Indian (2 percent), and Asian (1 percent).[50] In addition, Mariana, Jesus, and Callie's boyfriend AJ (Tom Williamson) are foster children of color. However, in the three other Freeform shows with foster characters, Emma in *The Lying Game*, Caleb in *Pretty Little Liars*, and Ricky in *The Secret Life of the American Teenager*, are all Caucasian.

The Fosters further addresses the cycle of foster care, where alumni of the foster care system face an increased likelihood that their children will be put in foster care as well. Daphne (Daffany McGaray Clark), a girl who was in both juvenile detention and Girls United at the same time as Callie, was placed into foster care because her mother was unable to care for her, and Daphne later became involved in gangs and drugs. Daphne then lost custody of her daughter Tasha (Chare'ya Wright). Although she worked to get her daughter back, it took a significant amount of time, during which Tasha became attached to her foster parents and no longer recognized Daphne as her mother.

Because of the scarcity of the foster portrayals, getting it right is of great importance. Interviews with the creators of *The Fosters* demonstrate the care they take in trying to be accurate as possible. As contributors to the Adopt Us Kids website point out, however, *The Fosters* sometimes glosses over the realities of foster care placement.[51] For example, the show does not adequately address rules for fostering (e.g., number of beds required, reporting incidents to social services) or the work involved in addressing behavioral issues. Neda Ulaby's focus group of former foster children criticized Callie's "bureaucracy-free entrance into the family."[52] According to the show's timeline, it was likely that Stef and Lena's certification to foster had lapsed when Callie was initially placed in their home.[53] Mike's (Danny Nucci) certification to foster AJ was pushed through very quickly despite the 30-hour training and home visits necessary in real life.[54]

At times, the red tape of foster care is glossed over to save screen time because getting certified to foster has little dramatic tension.[55] It is much more romantic to see Caleb and Hanna from *Pretty Little Liars* roast marshmallows over an open fire outside the tent he is living in after he ran away from his foster placement than focusing on his homelessness.[56] Critics fear, however, that unrealistic expectations or happily-ever-after fantasies will attract unqualified or unprepared individuals to foster. These fears may be alleviated by the fact that all potential foster parents are required to undergo training and complete a home study.[57] One mother, Molly Miller Rice, stated, "It's not exactly realistic…. But I am glad that they are putting foster care and adoption from foster care in the national spotlight."[58]

The Social Workers

Catherine Hiersteiner noted that social workers are likely misrepresented as either "young fallen angels" or "older spinsters or misguided mother types."[59] According to Josephine Nieves, "little troubles professional social workers more than the less-than-accurate image the public seems to have of our profession, acquired unfairly and based on misinformation."[60] Social workers are either largely absent or ineffective in Freeform portrayals of foster care.

In *The Fosters*, Callie's initial case worker Bill (Geoffrey Rivas) is overburdened but compassionate and empathetic (e.g., he describes Callie and Jude as "really great kids" who have had bad luck).[61] While Margaret is portrayed as a compassionate foster mother to Ricky in *The Secret Life of the American Teenager*, the audience does not see her much in her official role as a social worker. Caleb's social workers are unhelpful and untrustworthy (*Pretty Little Liars*). Caleb refers to his social worker, the fourth one in six months, as "a joke."[62] He does not report his neglectful foster parents because he does not trust his social worker. Foster children's distrust of social workers continues even in cases of sexual abuse. After being sexually assaulted by her former foster brother Liam (Brandon W. Jones), Callie explained she is reticent to tell her social worker because "once you get tagged as 'sexually volatile,' you won't get placed."[63]

Whenever social workers or lawyers are involved, it is rarely an uplifting scene. Rather, they are portrayed as antagonists out to make life more difficult for foster kids rather than as acting in the children's best interests. For example, social workers did little to protect Callie from her previous foster brother Liam's predatory sexual advances yet social services removed her from her safe home with the Adams-Fosters when they discovered that Stef and Lena's license to foster has expired.[64] Other than Stef and Lena, the Freeform shows mostly focus on the bad foster parents who indirectly reflect poorly on social workers since they are responsible for placing the children in the care of these ill-equipped and sometimes malicious foster parents. For example, Jude, Kiara (Cherinda Kincherlow), and Jack were allowed to slip through the cracks and remain in their respective abusive homes (*The Fosters*).

The risks faced by those working in the system may dissuade individuals from getting involved. Rita (Rosie O'Donnell), head of the Girls United group home, is compassionate toward the girls under her care there, but the residents often reject her support. When she tries to save one of the girls who ran away from the group home and was working as a prostitute, Kiara spat in Rita's face, saying she did not need help.[65] Another resident of the group home falsely accused Rita of hitting her.[66] Callie tricks Carmen (Alicia Sixtos) into retracting her accusation, but not before Rita was taken away in handcuffs. Even supportive foster parents can be victimized as foster teens experience frustration. When she first arrives at Stef and Lena's home, Callie refers to them as "dykes." Stef, while trying to be sympathetic, describes Callie as "not the warmest kid she has ever met."[67] Additionally, anger displacement occurs in *The Secret Life of the American Teenager* after Sanjay calls Ricky "son" and Ricky sharply reminds Sanjay that he is not his father.[68]

The Foster Parents

Related to the misrepresentations of social workers as sinners or saints, foster parents are similarly portrayed as one extreme or the other. Negative stereotypes of greedy foster parents may cause good people to fear being seen as fostering only for the money. The president and CEO of the Entertainment Industries Council, Bryan Dyak, appealed to producers to consider portraying foster parents as caring and supportive rather than only as mean or malicious.[69] While showing foster parents as greedy and abusive to the children for whom they care is damaging to recruitment, the saint-like perceptions of foster parenting are isolating and unattainable. These extremes are likely to dissuade individuals who think they are not good enough or that the work is too hard for "average people."

Stef and Lena on *The Fosters* represent a realistic view of foster parents: good-hearted individuals with flaws. Although their resources are strained, they vow to not send Callie and Jude back into the system. In the pilot episode of *The Fosters*, a social worker calls Lena to ask if they will take in a temporary placement since they had previously taken in and eventually adopted two children from the foster system. Lena is reluctant to bring someone dangerous into their home after the social worker tells her that Callie is in juvenile detention for damaging her foster father's property. After seeing Callie bruised and vulnerable, her eyes brimming with tears, however, Lena takes her home. Still, there are moments of second thoughts. Lena doubts they have enough room, but Stef reassures her. In a touching scene, Stef reassures Callie that she is neither disposable nor worthless.[70] They also make it clear to their adopted children and Stef's biological son that they love them equally. Stef's ex-husband Mike asks why they feel the need to "take in strays" like some "homeless shelter."[71] Mike's attitude evolves over time; he eventually becomes a foster parent himself. While not portraying their relationship with the same depth as *The Fosters*, *The Secret Life of the American Teenager* shows Ricky has been with his foster parents for years, and they love him like a son.

In contrast to Callie and Ricky's caring foster parents, Caleb explains that his foster parents do not care where he is, as long as they can cash the checks (*Pretty Little Liars*).⁷² They fail to report him missing when he runs away for fear of losing their income. In season two, it is mentioned that Caleb's foster mother Janet (Janet Borrus) continues to collect state money even while he lives elsewhere. She shows up at school to scold him for endangering her scam.⁷³ Similarly, when Clarice's foster license is revoked and she no longer receives money from the state after it is discovered that Emma is gone, she calls Emma an "ungrateful bitch" (*The Lying Game*).⁷⁴

In *The Fosters*, the cases get much darker in its portrayal of abusive foster parents. Before being placed with the Adams-Fosters, Callie took a bat to her foster father's car in order to stop him from beating her brother, Jude. Jim (Colby French), the abusive foster father, calls the police and has Callie arrested. He tells police that she went crazy and he was defending himself. The system appears blind to abusive foster parents as the authorities side with Jim and place Callie in a juvenile detention center, where she starts the series. When Callie comes back to rescue Jude, Jim points a gun at her.⁷⁵ He is eventually subdued and arrested. After his release, Jim gets yet another placement, a boy whom he later beats to death.⁷⁶ In another example of dangerous foster homes, the police arrived at former Girls United member Kiara's foster home after a 911 call and witnessed a man with a knife.⁷⁷ Despite the incident, Kiara was to be sent back to this abusive home. She later ran away to live on the streets and survived through prostitution.

Inept foster parents are also portrayed as failing to protect their placements from their biological children who take advantage of their foster siblings' vulnerable status. At a previous placement, Callie's foster brother, Liam, sexually victimized her. Just when the audience thinks Liam's mother is going to apologize to Callie, she sides with her biological son and tells Callie she wishes she had never met her.⁷⁸ Liam later targets another girl placed with his family. While less extreme, Emma's foster brother spies on her through the window when she is changing (*The Lying Game*). He also opens her bedroom door without knocking. In another scene, he grabs her by the wrists and tells her she is turning him on. He tries kissing without her consent. She calls him a "pervert" but he just laughs. His mother sides with him.⁷⁹ Significantly, ABC Family/Freeform has shown a range of foster parenting, illustrating the complicated reality.

The Birth Parents

The teens in these series are placed in foster care for reasons ranging from neglect to abandonment due to drugs to sexual abuse perpetuated by the birth parents. There is a paucity of compassionate stories about birth parents who fell on hard times, although across the country in 2015, 14 percent of foster kids (37,243) entered the system because of a caretaker's inability to cope and 10 percent (27,002) entered because of inadequate housing.[80] Gaylynn Burroughs, a family defense attorney, reported "at least 60 percent of child-welfare cases in the United States involve solely allegations of neglect, usually for inadequate food, clothing, shelter or inadequate supervision or guardianship."[81] One exception to the abuse portrayals is in *The Fosters,* where AJ's grandmother took custody of AJ and his brother after their father died and their mother was committed to a mental institution. The grandmother now suffers from dementia and lives in a nursing home. AJ and his brother end up on the street fearing they will be put into the system, which AJ eventually is.

The portrayals of birth parents are harsh but often redemptive over time. All of the shows included in this analysis have storylines in which the biological parents reemerge after a number of years to seek forgiveness or exploit their children further. Jesus and Mariana's mother, Ana (Alexandra Barreto), who chose getting high over caring for her five-year-old twins, leaves them at a police station.[82] When she returns ten years later, it is to ask Mariana for drug money.[83] Ana eventually gets sober and reestablishes a relationship with her children; however, she refuses to acknowledge the extent of her neglect when Mariana confronts her about it in therapy.[84] Callie and Jude's father Donald (Jamie McShane) returns after being released from prison to make amends with his children when Stef and Lena request he terminate his parental rights so they could adopt Callie and Jude.[85] Callie discovers that Donald is not her biological father.[86] Her biological father Robert (Kerr Smith) shows up later, not knowing he had a daughter. He tries to force Callie to be part of his family and blocks her adoption to Stef and Lena. He eventually signs the papers after seeing how committed Callie is to being adopted.

In another instance, *Pretty Little Liars'* Caleb disclosed that his biological mother Claudia (who is never shown on screen) dropped him off at his aunt's house when he was five years old.[87] When Claudia did not return, he was sent to live in multiple foster homes. Twelve years after she abandoned him, Claudia hired a private investigator to find Caleb. After speaking on the phone, Caleb flies to cross country to see her.[88] He ultimately forgives her for abandoning him but returns to Rosewood to be with Hanna. In *The Secret Life the American Teenager*, Ricky's biological father Bob (Bryan Callen) returns and blames his son Ricky for being in foster care, claiming Ricky lied about Bob sexually abusing him. Ricky's mother Nora (Anne Ramsey) was addicted to drugs and did nothing to protect Ricky from his abusive father Bob.[89] After Bob is rearrested, Ricky seeks answers from his mother Nora about her role in enabling the molestation.[90] Ricky expresses resentment that she never tried to protect him, but eventually forgives her.

Lastly, Emma and Sutton's biological mother gave them up for adoption because their father was married to someone else (*The Lying Game*). In a complex plot, a family friend sent Sutton to live with their biological father, but Emma was placed with a mentally unstable woman and sent to foster care at age three. Sutton finds their biological mother, Rebecca (Charisma Carpenter), who claims she loves them and never wanted the twins to be separated.[91] She wanted them in a nice home, with a nice family, because she was not capable of raising them. Emma unleashes her anger about being placed in foster care. Although Rebecca apologizes, she plots to break up the twins' biological father's marriage to reunite the family. Sutton warns Emma not to go soft and fall under Rebecca's spell. Unfortunately, the series was canceled before there was a resolution. By allowing birth parents a presence, ABC Family/Freeform allows a broader narrative to be created about foster care.

The Foster Teens

Unlike inclusion of foster children in television dramas as a one-off plot point, foster care is woven throughout the storylines and the characters on Freeform. In the Freeform shows, no character is solely defined by her or his foster status. When characters are singularly defined as "foster," it creates an "otherness" from which children—and characters—cannot escape. Instead of portraying foster children as victims or damaged perpetrators of crimes, for example, the characters have agency over their destinies. They are portrayed with such nuance that the audience can clearly understand their actions and incentives. The Freeform portrayals of foster care are powerful because of their emic and etic approaches. Calling attention to the difficulties children raised in foster care experience is commendable, but if success stories are not shown, it leaves children represented as damaged victims. The shows are educational to outsiders with little experience with the foster system and empowering, positive, and realistic for kids who have been in foster care. Thus, the shows not only allow outsiders a view into life in foster care, they also present positive and realistic reflections of life in foster care for people in the system.

Chmielewski commends the way Callie is portrayed.[92] In Chmielewski's personal experiences in the system, "foster youth are subdued, frightened and over-observant." Rather than blurt out what they think and feel, they are likely to keep it in. They are "closed off, untrusting and damaged."[93] He sees *The Fosters* in contrast to shows that feature one-time appearances of former foster children who are victims or perpetrators of crime or are weakly developed characters in a storyline involving one of the main characters who temporarily take in a young foster child such as on *Chicago Fire* (NBC, 2012–present)[94] or *Law & Order: SVU* (NBC, 1999–present).[95] By contrast, *The Fosters* chooses to let the story unfold at a more natural pace. This pacing creates an entertaining though educational setting through which *The Fosters* can teach outsiders about the foster system. More importantly, however, it adds realism to the show for those who have personally experienced foster care. The ups and downs of the storylines more accurately reflect experiences in the system, and thus these characters can be role models for foster kids.

The EIC warned that the media too often portrays children in the foster system as having severe behavioral issues.[96] Instead, EIC president Dyak suggests that when they are shown acting out, it is not because there are bad or damaged, but simply because they are teenagers—similar to children not in the system. In *The Fosters*, all five Adams-Foster children—biological, adopted, and foster—get into trouble at school or with the law, but their current or former foster status is never blamed.

Trust is another issue with which foster children struggle. Because Emma was placed with six foster families before the Mercers, she explains that she has learned not to get attached to people (*The Lying Game*). When Hanna offers Caleb the option to stay at her house (*Pretty Little Liars*), he declines, saying that her mother may start asking questions and report him for running away. Caleb eventually accepts her help after finding his belongings in the garbage at school where he had been sleeping. Caleb is relegated to a basement couch, hiding from Hanna's disapproving mother, Ashley (Laura Leighton), who had refused to let Caleb stay in the guest room. Eventually, Caleb earns her trust, and Ashley admits that she was wrong about him. Wary of being kicked out of the system when he turns 18 in a year, Ricky asks to be an emancipated minor so he leave on his terms and so another foster kid can receive the same support from his foster parents that he did (*The Secret Life of the American Teenager*).[97]

In addition to all the challenges with which they must cope, foster kids often experience prejudice from outsiders. Dyak warned that "there is a preconceived notion that if birth parents were drug addicts or had behavioral issues that their children will also act out."[98] Celeste Bodner, Executive Director of the Foster Club, accused the media of reporting prejudiced, negative information about children and young adults in foster care when "in reality, foster children and foster youth across America are waging courageous struggles to overcome tragic circumstances in their lives."[99] Foster children have to endure stigmatization from their peers, as well as adults such as teachers, principals, and parents. Sometimes adults stigmatize foster children through sympathy, extra attention, and kindness. At other times, the stigma manifests itself as distrust and desire to not expose "normal" children to the potentially disturbed foster children who are assumed to be bad influences. At her new school, the other students shunned Callie at the lunch table.[100] When someone is accused of selling drugs at school soon after she started, actual drug user Kelsey (Anne Winters) falsely accuses Callie of selling them because she knows Callie had just gotten out of juvenile detention.[101] The Girls United storyline also included the residents holding an open house so that suspicious neighbors could see the positive attributes of the home and allay their fears.[102] In *The Secret Life of the American Teenager*, Ricky fears that if his girlfriend Grace's (Megan Park) parents discover that his father is in prison for sexually abusing him, they will judge him and not let him see her. In this case, Grace and her family assure him that the abuse was not his fault.[103]

Socioeconomic issues add to stigma of foster care. Families with substantial resources are often able to get help for troubled parents (e.g., drug treatment) or arrange kinship placements, thus avoiding state-run foster care. In *The Lying Game*, appreciation of family and material comfort highlights class differences between Emma and Sutton. Sutton, the rich twin, disdains her family, while Emma, who grew up in foster homes, calls the family "truly a gift."[104] In *Pretty Little Liars*, Caleb, a "street-smart kid" with a "sketchy past," explains he used to work for car thieves as a computer hacker to survive.[105] After his father was sent to prison, Ricky was placed in foster care because his mother was homeless and addicted to drugs (*The Secret Life of the American Teenager*).[106] Jude's down-and-out father Donald is contrasted with Callie's wealthy birth father Robert (*The Fosters*).[107]

On Freeform, children in foster care are also portrayed positively, gaining self-confidence and self-empowerment, and the teens are active in their development. In *The Secret Life of the American Teenager*, Ricky tells his foster parents that he will leave so "the next Ricky" (i.e., another abused foster child) can come into the home. At one point in *The Fosters*, AJ is tempted to run away with his brother but decides his best chance at life is to stay in his new foster home. Seemingly bad characters can turn their lives around. Daphne assaulted Callie when they were in juvenile detention together but later expresses remorse, and the two become friends when they are both in Girls United (*The Fosters*). Daphne is shown trying to turn her life around (i.e., getting a job and apartment and staying out of trouble) so she can regain custody of her daughter.

These shows reassure the audience that, even after experiencing abuse and neglect, foster children can develop healthy relationships. Caleb tells Hanna that after living in foster care, his relationship with her has greatly changed his outlook on the future and that he felt alone his whole life until he met her.[108] Emma tells Sutton that while in foster care, all she wanted to do was run away, but since she started living Sutton's life, she finally stands up for herself and stays.[109]

Conclusion

According to Susan Hoag Badeau, "the media has an enormous opportunity to be catalysts for healing and hope for the half million children in America's foster care system."[110] Realistic portrayals "offer compelling drama, enjoyable entertainment and valuable education to the public."[111] Portraying teens in foster care as resilient rather than damaged changes how they are seen to the outside world.[112] It follows that the more positive the stories, the more the public's perception will change.[113] In addition, simply normalizing foster care issues may also help attract potential foster parents. By offering more depictions of foster families on television, Americans may recognize themselves in those portrayals and see themselves as potential foster parents.[114] Engaging storylines have the ability to recruit families for children in need of loving homes.

The EIC encourages media producers to work with child welfare experts to improve and increase awareness of issues foster families face.[115] Programming can be entertaining and educational, and thus have a greater ability to influence society and effect social change with regard to foster care-related issues. Even if Freeform shows contain some factual inaccuracies leading to unrealistic expectations in order to appeal to television audiences, they still bring exposure to foster care.[116] Chmielewski proposed that *The Fosters* be essential viewing for all who are involved with foster care.[117] Erica Schaaf added that *The Fosters* was seen as a learning tool and, thus, social workers were encouraged to watch the show.[118] The shows on Freeform do an excellent job of educating viewers by focusing on and showing the world through the lens of the foster children. Because of the young adult-driven narrative, the audience sees the frustrations, joys, and struggles of foster care through the eyes of someone in it, creating an engaging educational environment wherein people without direct experience of foster care could learn more and those in the system could be empowered through their representations.

The Fosters, and by extension Freeform, is an important vehicle for telling the story of foster care. The cable network's slogan, "A New Kind of Family," is at the heart of shows such as *The Fosters,* showing at any type or configuration of family—as long as it is loving and safe—is a good one. *The Fosters* illustrates that "it's okay to be in foster care, and it's okay to be foster parents."[119] When Brandon explains to Callie that Stef and Lena asked if it was okay for them to take in the twins as foster children, Callie is perplexed why he said yes.

> BRANDON: "I figured there was enough to go around."
> CALLIE: "Enough of what?"
> BRANDON: "Everything."[120]

NOTES

1. Julie Prince, "'Family Friendly?!' Disney's ABC Family Channel 'The Fosters' Airs Youngest Gay Kiss," *Clash Daily*, March 6, 2015, http://clashdaily.com/2015/03/family-friendly-disneys-abc-family-channel-the-fosters-airs-youngest-gay-kiss/.

2. Laura Bradley, "ABC Family Is Smart to Change Its Name. Maybe Other Networks Should Follow Suit, Too," *Slate Magazine*, October 6, 2015, http://www.slate.com/blogs/browbeat/2015/10/06/abc_family_s_name_change_to_freeform_is_smart.html.

3. Herb Scribner, "By Changing Its Name, ABC Family Clarifies It's Not Family Friendly," *Deseret News*, October 07, 2015, http://national.deseretnews.com/article/6321/by-changing-its-name-abc-family-clarifies-its-not-family-friendly.html.

4. *Ibid.*

5. *Ibid.*

6. Brian Tashman, "One Million Moms Protests 'The Fosters' for Trying to Desensitize America and Our Children by Promoting Inappropriate Behavior," *Right Wing Watch*, June 7, 2013, http://www.rightwingwatch.org/content/one-million-moms-protests-fosters-trying-desensitize-america-and-our-children-promoting-inap.

7. *Ibid.*

8. Tavirn Peart, "Jennifer Lopez's New Show 'The Fosters' Stirs Up Controversy Over Same-Sex Couple Raising Family," *Latinos Post*, April 15, 2013, http://www.latinospost.com/articles/16674/20130415/jennifer-lopezs-new-show-fosters-stirs-up-controversy-over-same.htm.

9. Tashman, "One Million Moms."

10. David Bauder, "'The Fosters' Credits Negative Attention from One Million Moms for Strong Season," *LGBTQ Nation*, January 17, 2014, http://www.lgbtqnation.com/2014/01/the-fosters-credits-negative-attention-from-one-million-moms-for-strong-season/.

11. Rich Valenza, "Bradley Bredeweg, Executive Producer, Discusses ABC Family's 'The Fosters,'" *Huffington Post*, January 10, 2014, http://www.huffingtonpost.com/2014/01/10/bradley-bredeweg-the-fosters_n_4569014.html.

12. Gretchen Livingston, "Fewer than Half of U.S. Kids Today Live in a 'Traditional Family,'" *Pew Research Center*, December 22, 2014, http://www.pewresearch.org/fact-tank/2014/12/22/less-than-half-of-u-s-kids-today-live-in-a-traditional-family/.

13. "Foster Care," *Children's Rights*, http://www.childrensrights.org/newsroom/fact-sheets/foster-care/.

14. *Ibid.*

15. Teresa Wiltz, "Drug-Addiction Epidemic Creates Crisis in Foster Care," *The Pew Charitable Trusts*, October 7, 2016, http://www.pewtrusts.org/en/research-and-analysis/blogs/stateline/2016/10/07/drug-addiction-epidemic-creates-crisis-in-foster-care.

16. *Ibid.*

17. Serge Moscovici, *Social Representations: Explorations in Social Psychology* (Hoboken: Wiley, 2001), 37.

18. *Ibid.*, 27.

19. Michael Murray, "Connecting Narrative and Social Representation Theory in Health Research," *Social Science Information*, 41, no. 4 (2002): 656.

20. Damien W. Riggs, Daniel King, Paul H. Delfabbro, and Martha Augoustinos, "Children Out of Place: Representations of Foster Care in the Australian News Media," *Journal of Children and Media* 3, no. 3 (2009): 234–248, 235.

21. *Ibid.*, 236.

22. "Safe Haven." *Criminal Minds*, season 6, episode 5, CBS, October 20, 2010.

23. "Born Bad." *Law & Order*, season 4, episode 9, NBC, November 16, 1993.

24. Riggs et al., "Children Out," 242.

25. Entertainment Industries Council, "Picture This: A Resource for Media Creators and News Media," 2009, http://www.eionline.org/resources/publications/z_picturethis/Foster_Care_FINAL.pdf.

26. Riggs et al., "Children Out," 244.

27. Andy West, "They Make Us Out to be Monsters: Images of Children and Young

People in Care," in *Social Policy, the Media and Misrepresentation*, ed. Bob Franklin (London: Routledge, 1999): 253–267, 253.
28. Riggs et al., "Children Out," 245.
29. Lloyd Nelson, "Foster Parent Shortages in the United States," *Foster Care Newsletter*, March 1, 2016, http://foster-care-newsletter.com/foster-parent-shortages-in-the-united-states/#.WFILSLIrKM8.
30. *Ibid.*
31. Wiltz, "Drug-Addiction."
32. Entertainment Industries Council, "Picture This."
33. *Ibid.*, 9.
34. *Ibid.*, 6.
35. Chris Chmielewski, "A Special Review of: The Fosters," *Foster Focus Magazine* 2, no. 12 (2013), http://www.fosterfocusmag.com/articles/special-review-fosters (emphasis added).
36. U.S. Department of Health and Human Services, Administration for Children and Families, Administration on Children, Youth and Families, Children's Bureau, "The AFCARS Report," 2015, https://www.acf.hhs.gov/sites/default/files/cb/afcarsreport22.pdf.
37. Neda Ulaby, "Foster Families Take Center Stage," *NPR*, June 3, 2013, http://www.npr.org/sections/monkeysee/2013/06/03/188326453/foster-families-take-center-stage.
38. "Pilot," *The Lying Game*, season 1, episode 1, ABC Family, August 15, 2011.
39. "Sibling Issues in Foster Care and Adoption." *Child Welfare Information Gateway*, January 2013, https://www.childwelfare.gov/pubPDFs/siblingissues.pdf#page=1&view=Introduction.
40. Emily Nussbaum, "Sweet and Low: 'The Fosters' and 'Broad City,'" *The New Yorker*, February 10, 2014, http://www.newyorker.com/magazine/2014/02/10/sweet-and-low-2.
41. Valenza, "Bradley Bredeweg."
42. *Ibid.*
43. *Ibid.*
44. Ulaby, "Foster Families."
45. Erica Schaaf, "Exclusive Interview with The Fosters' Cierra Ramirez and Bradley Bredeweg," *Talk Nerdy with U.S.*, March 21, 2016, http://talknerdywithus.com/2016/03/21/exclusive-interview-with-the-fosters-cierra-ramirez-and-bradley-bredeweg/.
46. Ulaby, "Foster Families."
47. "Pilot," *The Fosters*, season 1, episode 1, ABC Family, June 3, 2013.
48. "The Show," *The Fosters*, season 3, episode 19, Freeform, March 21, 2016.
49. "Kingdom Come," *The Fosters*, season 3, episode 20, Freeform, March 28, 2016.
50. Adopt Us Kids, "News and Announcements," accessed February 01, 2017, http://adoptuskids.org/news-and-announcements/in-the-news/story?k=foster-care-on-the-fosters
51. *Ibid.*
52. Ulaby, "Foster Families."
53. "Pilot," *The Fosters*.
54. "Idyllwild," *The Fosters*, season 3, episode 9, ABC Family, August 10, 2015.
55. Ulaby, "Foster Families."
56. "A Person of Interest," *Pretty Little Liars*, season 1, episode 19, ABC Family, February 28, 2011.
57. Adopt Us Kids, "News."
58. *Ibid.*
59. Catherine Hiersteiner, "Saints or Sinners? The Image of Social Workers from American Stage and Cinema Before World War II," *Affilia: Journal of Women and Social Work* 13 no. 3 (1998): 14–26. Quoted in Deborah P. Valentine and Miriam Freeman, "Film Portrayals of Social Workers Doing Child Welfare Work," *Child and Adolescent Social Work Journal* 19, no. 6 (2002): 456.
60. Josephine Nieves, "Building a Better Public Image," *NASW News* (May 1997): 2. Quoted in Deborah P. Valentine and Miriam Freeman, "Film Portrayals of Social Workers Doing Child Welfare Work," *Child and Adolescent Social Work Journal* 19, no. 6 (2002): 456.
61. "Consequently," *The Fosters*, season 1, episode 2, ABC Family, June 10, 2013.

62. "Je Suis Une Amie," Pretty Little Liars, season 1, episode 16, ABC Family, February 7, 2011.

63. Nussbaum, "Sweet and Low."

64. "Things Unknown," The Fosters, season 2, episode 1, ABC Family, June 16, 2014.

65. "Now Hear This," The Fosters, season 2, episode 18, ABC Family, March 2, 2015.

66. "Daughters," The Fosters, season 3, episode 8, ABC Family, August 3, 2015.

67. "Pilot," The Fosters.

68. "Absent," The Secret Life of the American Teenager, ABC Family, season 1, episode 7, August 12, 2008.

69. Entertainment Industries Council, "Picture This," 4.

70. "Pilot," The Fosters.

71. Ibid.

72. "Je Suis Une Amie," Pretty Little Liars.

73. "The Devil You Know," Pretty Little Liars, season 2, episode 5, ABC Family, July 12, 2011.

74. "Sex, Lies and Hard Knocks High," The Lying Game, season 1, episode 9, ABC Family, October 10, 2011.

75. "Consequently," The Fosters.

76. "The Show," The Fosters.

77. "Stay," The Fosters, season 2, episode 13, ABC Family, January 26, 2015.

78. "Now and Then," The Fosters, season 4, episode 4, Freeform, July 18, 2016.

79. "Never Have I Ever," The Lying Game, season 1, episode 8, ABC Family, October 3, 2011.

80. U.S. Department of Health and Human Service, "The AFCARS Report."

81. Gaylynn Burroughs, "Too Poor to Parent?" The Huffington Post, November 17, 2011, http://www.huffingtonpost.com/gaylynn-burroughs/too-poor-to-parent_b_109971.html.

82. "Vigil," The Fosters, season 1, episode 9, ABC Family, July 29, 2013.

83. "Pilot," The Fosters.

84. "Until Tomorrow," The Fosters, season 4, episode 20, Freeform, April 11, 2017.

85. "Don't Let Go," The Fosters, season 1, episode 19, ABC Family, March 10, 2014.

86. "Things Unknown," The Fosters.

87. "A Person of Interest," Pretty Little Liars.

88. "Picture This," Pretty Little Liars, season 2, episode 9, ABC Family, August 9, 2011.

89. "The Father and the Son," The Secret Life the American Teenager, season 1, episode 14, ABC Family, October 21, 2009.

90. "Choices," The Secret Life the American Teenager, season 2, episode 21, ABC Family, March 1, 2010.

91. "To Lie For," The Lying Game, season 2, episode 10, ABC Family, March 12, 2013.

92. Chmielewski, "A Special Review."

93. Ibid.

94. Chicago Fire, season 5, NBC, 2016–2017.

95. Law & Order: SVU, season 12, NBC, 2010.

96. Entertainment Industries Council, "Picture This," 10.

97. "Born Free," The Secret Life of the American Teenage, season 2, episode 5, ABC Family, July 20, 2009.

98. Entertainment Industries Council, "Picture This," 10.

99. Ibid., 2.

100. "Pilot," The Fosters.

101. "Consequently," The Fosters.

102. "Girls Reunited," The Foster, season 2, episode 8, ABC Family, August 4, 2014.

103. "That's Enough of That," The Secret Life of the American Teenager, season 1, episode 15, ABC Family, January 26, 2009.

104. "Over Exposed," The Lying Game, season 1, episode 5, ABC Family, September 12, 2011.

105. "Surface Tension," Pretty Little Liars, season 2, episode 7, ABC Family, July 26, 2011.

106. "Absent," *The Secret Life of the American Teenager*, season 1, episode 7, ABC Family, August 12, 2008.

107. "Father's Day," *The Fosters*, season 3, episode 2, ABC Family, June 15, 2015.

108. "It's Alive," *Pretty Little Liars*, season 2, episode 1, ABC Family, June 14, 2011.

109. "To Live For," *The Lying Game*, season 2, episode 10, ABC Family, March 12, 2013.

110. Entertainment Industries Council, "Picture This," 11.

111. *Ibid.*, 31.

112. *Ibid.*, 10

113. *Ibid.*, 32.

114. *Ibid.*

115. *Ibid.*, 4.

116. Adopt Us Kids, "News."

117. Chmielewski, "A Special Review."

118. Schaaf, "Exclusive Interview."

119. Ulaby, "Foster Families."

120. "Pilot," *The Fosters*.

Deaf Cultural Values in *Switched at Birth*

SHARON L. PAJKA

Academics often trivialize the benefits of television; yet, it was television and the positive portrayals of deaf characters that completely transformed a recent conference experience for me. Last June, as I was mingling with some presenters, one noticed Gallaudet University listed on my badge and squealed, "Do you know Nyle DiMarco?!! Do you know the cast from *Switched at Birth*?!!" This was a vastly different experience than most conference introductions, when I have had to explain that Gallaudet University is the only liberal arts university in the world for Deaf and Hard-of-Hearing students, and that I, indeed, teach my courses through American Sign Language.

According to ABC Family, *Switched at Birth* (2011–2017) is "the first mainstream television series to have multiple deaf and hard-of-hearing series regulars and scenes shot entirely in American Sign Language (ASL)."[1] The series comes at a time when deaf characters and deaf actors are being included in theater, film, and television with more frequency, from Deaf West's *Spring Awakening* on Broadway and their performance at the Tony Awards (2016); to Nyle DiMarco winning both *America's Next Top Model* (UPN, The CW, VH1; 2003–present) in 2015 and *Dancing with the Stars* (ABC, 2005–present) in 2016; to Treshelle Edmond performing the National Anthem in ASL at the Super Bowl (2015). This recent increase in visibility of deaf performers derives from an increased interest in depicting diversity, a growing acceptance of those with disabilities on television, and a great deal of collective effort by members of the Deaf community.[2]

Switched at Birth is a family drama television series that premiered on ABC Family in 2011. It focuses on two teenage girls who were switched in the hospital after they were born, and subsequently were raised in different environments: Bay Kennish (Vanessa Marano) grows up in an affluent suburb of

Mission Hills, Kansas, raised by wealthy white parents, Kathryn and John Kennish (Lea Thompson and D.W. Moffett); while Daphne Vasquez (Katie Leclerc) grows up in the working-class neighborhood of East Riverside raised by a Latina single mother, Regina Vasquez (Constance Marie), who works as a hairdresser. The biological father, Angelo Sorrento (Gilles Marini), who is of French and Italian decent, abandons his family, believing that Daphne is not his child. When the two families come together to raise the girls, all of their lives become quite complicated. They discover how different their lives have been. In addition to the new family dynamics, the Kennish family learns that Daphne, who contracted meningitis at the age of three, is deaf. Daphne is a culturally Deaf person who attends a residential school for the Deaf with teachers who are Deaf, socializes with Deaf friends, and communicates by using ASL. The members of the Kennish family learn about Daphne, and they learn about Deaf culture and the Deaf community to which Daphne belongs. To do this, all of the characters learn some ASL.

The series includes numerous Deaf characters. However, an increase in characters does not necessarily translate to an increase in understanding of those members of a culturally Deaf community, nor does it translate into the most accurate, respectable, well-rounded characterization of the deaf. Historically, deafness has been used as a literary device to relay messages about the struggles of humankind and elicit sympathy, as metaphors for feelings of isolation, and as comedic devices for misunderstandings. Deaf characters have been included to further the plot or to help the audience understand the plight of a hearing character.[3] In film and television, "the symbol of success still remains speech."[4]

The latest technology allows filming close-ups to become more common and requires all actors, both deaf and hearing, to change their approach to the art.[5] These filming techniques can exclude the Deaf "voice" by removing ASL from the frames. In addition, the series falls into some of the same traps as past productions with deaf characters by situating itself within the mindset of the dominant culture offering a subtext of value for speech and hearing. Yet, *Switched at Birth* still pushes against the grain in many areas incorporating topics and issues that feature Deaf cultural values as well as numerous Deaf actors.[6]

ABC Family/Freeform delivers series with diverse and unique families that it believes its target audience finds "relatable."[7] The network spends a great deal of effort incorporating cultural values in order to engage viewers both during episodes and in social media while anticipating the next episodes. Through an analysis of the series and focusing upon two episodes including dream sequences, *Switched at Birth* demonstrates how including cultural values, and how working with Deaf actors, filmmakers, and writers can create a more realistic depiction of deafness and Deaf people.

Views of Deafness and Deaf People

Along with numerous nominations and awards, ABC Family made television history airing the special episode, "Uprising,"[8] shown almost entirely in ASL with subtitles for those viewers not fluent in the language.[9] By removing the sound and requiring viewers not fluent in ASL to read subtitles during this episode, this was the first time an episode of a mainstream show gave precedence to the visual nature of ASL over the auditory communication of the majority of television viewers. The network has helped increase exposure to issues related to deafness, and it appears to have helped advance the hiring of deaf actors by including reoccurring main and secondary characters along with special guest stars.

There are two distinct perspectives regarding deafness: the pathological and the cultural. In the medical or pathological perspective, deafness is a biological disability and a deviation from the "normal," healthy ability to hear. The emphasis focuses upon the physical condition of deafness versus the deaf person's life and abilities.[10] Proponents of the pathological perspective encourage deaf people into lipreading and speech therapy so that they can be more like members of the dominant culture and fit within societal norms. In contrast, the cultural perspective offers an acceptable term "Deaf" to refer to individuals in a cultural and linguistic minority group regardless of their actual degree of hearing loss. From a cultural perspective, hearing loss can be usurped to "Deaf Gain," a way of acknowledging and celebrating the contributions of Deaf people and its culture.[11]

Members of the Deaf community value their language, ASL, and hold various cultural norms and beliefs which differ from the dominant culture.[12] Deafness becomes an identity when it includes individuals participating in a community with its own clubs, churches, theatre companies, schools, sports organizations, and beauty pageants.[13] The cultural perspective does not deny hardships which occur from being deaf; however, the emphasis focuses upon overcoming language and communication barriers, often imposed by the dominant culture, rather than dealing with their deafness.[14]

The Deaf have had difficulty finding themselves in literature, film, media, or among other cultural representations of society. They have not been included frequently in mainstream media; however, when they are included, "selective images based on the medical model" have been the focus, showing "deaf people as subnormal."[15] Viewers of *Switched at Birth* have learned Deaf people can view themselves as members of a linguistic minority group, and that they can do everything except hear. The network's focus on diversity enables the series to present another way of showing the human experience.

Brand Management

In her article "Cultural Diversity as Brand Management in Cable Television," Melanie E.S. Kohnen explores how cultural diversity is used as a strategy for ABC Family's "new kind of family."[16] She notes that the network's research shows that millennials appreciate family-oriented television "when it transcends the genre's stereotypes."[17] She explains that this imagined audience values family but not necessarily the "nuclear, white, Christian family of television's past," but those which include diversity.[18] Kohnen argues that there is "a more profound engagement with diversity [in *The Fosters*] where racial, ethnic, and sexual difference is openly addressed," unlike channels and shows with more managed diversity, which avoid social critique.[19] She lists *Switched at Birth* as another example of a show that openly addresses racial, ethnic, and sexual difference.[20]

ABC Family programs include active media sites encouraging audience members to interact with their shows. Kohnen notes that diversity is central to the social media interaction. For *Switched at Birth*, understanding diversity extends to Deaf cultural values. While the series includes characters that are diverse in race, ethnicity, and sexual orientation, and includes episodes that address intersectionality, the show's focus on diversity remains with deafness.[21]

ABC Family knows that their audience members value diversity. *Switched at Birth* was the network's highest-rated series debut of all time, attracting 3.3 million viewers.[22] It is not surprising that the network continues to spend a great deal of time and effort on media sites used with their series. Their efforts reinforce the show's interest in viewing deafness as a cultural perspective. For example, the network posts videos of the actors from the television series sharing an ASL word of the day accompanied by the hashtag #ASLLessonoftheDay on social media. These videos, which reinforce Deaf cultural values, include both Deaf and hearing actors, as well as deaf and hearing characters showing that it is everyone's responsibility to learn and share ASL.

Another popular hashtag, #TakeBackCarlton, was used during the episode "Uprising," when the students, both Deaf and hearing characters, take over the Carlton School for the Deaf in order to save it from being closed. Again, this shows that it is everyone's responsibility to preserve such cultural institutions. While Carlton School for the Deaf is a fictional school, the episode's focus demonstrates one of the cultural values of the Deaf community: the appreciation and value of schools for the Deaf. Since the Americans with Disabilities Act (ADA) and Section 504 of the Rehabilitation Act of 1973, federal law requires that Deaf students receive equal access and an equal opportunity to participate in public schools. In addition, the Individuals with Disabilities Education Act (IDEA) also affects Deaf children as the law requires public school systems to provide a "free, appropriate public education" to children who need special education or related services. While these laws were intended to support access for Deaf individuals, they have led to a great deal of confusion on the best education opportunities for Deaf children. Schools for the Deaf have been places that foster the culture, heritage, and language of Deaf people. Prior to these laws, Deaf children attended schools for the Deaf and bonded with other Deaf individuals like them; after these laws, it is not unusual to find a Deaf child alone in a public school receiving nearly all of his communication, academic and social, through an interpreter. In such situations, the Deaf child may not have any direct communication with his teacher, only indirect correspondence through an interpreter. Further, even on the playground an interpreter would be needed for the child to communicate with his peers.

The use of #TakeBackCarlton triggered social media and face-to-face conversations about the continued trend of state schools for the deaf being closed due to financial hardships. Similar to the Carlton protest, there have been numerous student-led protests to save schools for the Deaf. During the episode, Marlee Matlin's fictional character discusses her involvement with the actual Deaf President Now movement at Gallaudet University in 1988. The protest began on March 6, 1988, when the Gallaudet University Board of Trustees announced its decision to appoint a hearing woman who did not know ASL over other highly qualified Deaf candidates.

Finally, another important use of hashtags to take a stance about Deaf cultural values can be seen through the social media accounts from the actors involved with the show. One can infer that the network is in support of Deaf cultural values since one of the actors, Sean Berdy, uses "Support #DeafTalent" for his Twitter account that is linked to the network's Twitter. At the very least, they have not asked Berdy to remove the hashtag in order to distance himself from the movement.

History of Deaf Characters in Film and Television

In *Hollywood Speaks: Deafness and the Film Entertainment Industry*, John. S. Schuchman explains that many people have never met a deaf person, or if they have it is likely that due to communication barriers a meaningful dialogue did not take place. Although most have not met a deaf person, "most Americans have had an opportunity to see a deaf character depicted in a movie or television program."[23] Film and television have the power to shape our understanding and attitudes about groups of people. With deaf characters on television, viewers can see Deaf people in their homes; recently with social media, the audience now can have a dialogue with the Deaf or hearing actors who play deaf characters they watch on television.

Schuchman argues that in the past many of the deaf characters were portrayed in a way that perpetuated the "pathological condition" of deafness to the point that many people considered film and television "their nemesis in the struggle for increased equality in the twentieth century."[24] The inaccurate portrayal of deaf characters is not a new phenomenon. As far back as the early 1900s, Deaf community members protested films with hearing actors who inaccurately used sign language or invented signs.[25] During the silent movie era, deaf characters were depicted as helpless, victimized "objects of pity," or deafness was used as a trick by a hearing character in order to catch a rival.[26] Hollywood perpetuates the attitudes of the dominant culture. Until recently, it has been rare to see deaf characters represented as "normal" or desirable.[27]

In "Hollywood through Deaf Eyes: A Panel Discussion," actresses Linda Bove, best known for her role on *Sesame Street*, and Oscar-winning Marlee Matlin, who currently stars in *Switched at Birth*, explain how they each had to take on additional responsibilities in order to insure the most positive portrayals of their characters.[28] Bove explains: "First, I could advise the writers about problems in the scripts that should be changed. Years later, I found that I could also solve problems by functioning as a kind of director. I learned to instruct the directors on how to film me so that all the signs could be clearly seen. I had to work with the camera man to show the space needed to make sure signs were not cut out of frame."[29] Similarly, Matlin notes that she cannot wait for a deaf character role to come to her; she approaches studios to talk with scriptwriters and even suggests changing some characters that were intended to be hearing into deaf roles.

Deaf Actors

Deaf actors have been restricted to roles of deaf characters, although oftentimes deaf roles have been given to hearing actors. Even in the mid–1990s, Japanese actors who were deaf were prohibited from playing roles in film, television, or on stage.[30] To exacerbate tense conditions, Hollywood studios have asked for "crash-courses on deaf culture, sign language, deaf mannerisms, and the 'deaf style' of acting with no intention of hiring deaf actors for the deaf roles."[31] This frustration parallels the frustrations of actors with disabilities who continue to be displeased by Hollywood's choices of casting those without a disability in a role of a disabled character. S.E. Smith argues, "Nondisabled actors taking on disabled roles get critical acclaim and attention while their talented disabled counterparts, for the most part, are ignored."[32]

Representing an approach of community collectivism, Ann Silver states, "Those who succeed owe it to the deaf community to reach back and help others."[33] The current generation answers the call through the #DeafTalent movement. Actor Tyrone Giordano explains that the hashtag used in social media, which originally was used as a form of protest to call attention to deaf character roles being given to hearing actors, has "become a celebratory" one used to call attention to Deaf actors and deaf character roles.[34] Filmmaker Jules Damenon tracks the timeline of the #DeafTalent movement,[35] and Giordano has compiled an extensive database of talented deaf individuals, that he has made this available through their Facebook page.[36] Both the timeline and the database are intended to keep members of the Deaf community informed. In a collective effort to assist one another within the community, if an actor or producer needs to quickly find a representative from the Deaf community, the database is available for everyone to use. *Switched at Birth* features numerous Deaf actors and has showcased celebrities who are Deaf and Hard-of-Hearing, including T. L. Forsberg, a singer, songwriter, and actor; Ashley Fiolek, a retired professional motocross racer; Derrick Coleman, the first deaf offensive player in the NFL; and Nyle DiMarco, who as previously noted has become a reality show star.

Actress Katie Leclerc, who identifies as having a slight hearing loss from Ménière's disease, explains in an interview, "Within the deaf community you have to have a hearing loss to portray hearing loss and it's one of those political things that it's just the way it goes you know. If you're not a person with a hearing loss then you're taking the role away from someone who does." Leclerc is intentionally unclear as she explains the #DeafTalent movement in her interview. Using the term "hearing loss" instead of deaf allows her membership in the movement. As this author has Ménière's disease and experiences symptoms similar to Leclerc including hearing loss, Leclerc's statements misconstrue the Deaf experience. She has not spoken publicly about any discrimination she has felt based on having Ménière's disease. Because of this, she is perceived by some Deaf community members as coming from a place of privilege and passing. She explains that the nationwide casting call for Daphne looked for a deaf person to play the part.[37] Some viewers perceive Leclerc as someone taking advantage of a slight hearing loss to land the part, even noting that Leclerc required voice lessons in order to "simulate 'Deaf speech.'"[38] On March 30, 2017, Deaf community member Martha Anger used Facebook Live to share her displeasure over several issues within the Deaf community. In her video, she states that she worked on the set of *Switched at Birth* with Leclerc, and that Leclerc did not misconstrue her hearing loss but blatantly lied to the *Switched at Birth* producers about being deaf during her audition.[39] The character Leclerc plays has near flawless lipreading abilities, which is possible since the actress does not rely on lipreading skills.[40]

Wanting an accurate portrayal of a real deaf person and a script that includes the work of Deaf script writers is a shared preference for supporters of the #DeafTalent movement.[41] Deaf characters played by Deaf actors offer more authentic portrayals. S.E. Smith argues that there is an "ongoing problem of cripface and the tragedization of disability" in Hollywood when those without disabilities "are cast as disabled characters."[42] Seen as a marginalizing and disempowering practice, the term cripface refers to actors without disabilities playing characters with disabilities. Arguably, this parallels concerns of white people playing Black characters in films, but Smith emphasizes that those with disabilities playing characters with disabilities are not being recognized: "You win an Academy Award for being a nondisabled actor digging deep to play a disabled character."[43]

Recent Research on Deaf Characters in Film and Television

There have been few recent studies focusing on the portrayals of deaf characters in film and television. Overall, the conclusions point to inaccuracies in portrayals, such as Jennifer Rayman examining how ASL is visible or eliminated due to film editing decisions. Analyzing episodes of *CSI: New York* (CBS, 2004–2013) and *Law and Order: Criminal Intent* (NBC, USA Network; 2001–2011), she argues that the deaf characters that have ongoing roles still remain isolated from the Deaf community.[44] Through her analysis, she asserts that "the filmmakers use both sound and video editing techniques to mark the experiential difference between hearing and Deaf characters." Further, she explains that sound was muted or distorted, and extreme close-ups were used to focus in on the movement of lips or hands depending on the speaker. She emphasizes, "Though these techniques may heighten awareness of deaf experience to a non-signing audience they also point to a disabling stereotyping of the experience of being Deaf as lacking" and cites that this type of frame includes a pathological understanding of deafness.[45]

Building upon the work of Rayman, Miriam Lerner focuses her research on deafness and deaf characters' narrative function in film. Of the forty-six contemporary films she analyzed, each falls within ten classifications. Lerner explains that deafness is used as a plot device, a metaphor, a symbolic commentary on society, or a psychosomatic response to trauma. The deaf characters are protagonist informants or used as a parallel to the protagonist. Sign language has been used as the "hero" and as a way to express thoughts that the main hearing character is having but is unable to put into words. Finally, there are stories about deaf/hearing relationships and those with "a-normal-guy-or-gal-who-just-happens-to-be-deaf" such as the Deaf character in *The Family Stone*.[46] Arguably many of the Deaf characters played by Deaf actors on *Switched at Birth* are portrayed as "a-normal-guy-or-gal-who-just-happens-to-be-deaf." Emmett (Sean Berdy) enjoys art and works on his motorcycle; Travis (Ryan Lane) plays baseball. There is nothing extraordinary about these interests. The writers intentionally portray these characters as regular teens doing what regular teens enjoy doing.

Seon-Kyoung An et al. analyzed the effects of watching *Switched at Birth* on viewers' attitudes toward deafness.[47] Participants completed pretest questionnaires, watched episodes in the series, and responded to follow-up questionnaires after each episode. The research team concluded that the "viewers' attitudes toward deafness significantly improved overall."[48] However, they found that participants "felt more strongly after viewing [the episodes] that deaf people should prioritize learning to speak and lip read, even to the potential exclusion of sign language."[49] Their conclusions reveal that participants believe that deaf people need to try harder to communicate, placing the responsibility of communication on the deaf. As Schuchman explains, "deaf people only have difficulties when they are required to interact with the dominant cultures who hear."[50] The researchers of the study noted that the "Deaf protagonist Daphne navigates successfully through the hearing world in large part because she is able to speak and read lips."[51] In the episodes for their study, Daphne only communicates in ASL when she interacts with her mother and with Emmett.

Exploring fictional representations of hearing loss, Katherine A. Foss performed a textual analysis of 276 episodes of television shows that included storylines about deaf characters or deafness from the last few decades.[52] She argues, "As opposed to d/Deaf characters, who are always played by Deaf actors, all characters experiencing hearing loss in the program are played by hearing actors."[53] Actually, there are many roles for deaf characters that are given to hearing actors. As previously noted, the surge in the #DeafTalent movement calls attention to such incidents where deaf character roles are being filled by hearing actors. For example, filmmaker Michael Ojeda cast a hearing actress as a deaf rape victim in the film *Avenged*, insisting that there would be too many risks if the role was filled by a deaf person.[54] Another example is when Catalina Sandino Moreno was cast in the role of a deaf, mute mother in *Medeas*.[55] Even the horror movie *Hush*, which features a culturally Deaf character, is played by a hearing actress.[56]

Foss concludes that representations of deafness and Deaf culture on television reveal that the pathological model of deafness is still used with the deaf characters being shown as "vulnerable or dehumanized" with a search for a "'cure' for deafness" remaining central to the storyline.[57] However, she argues that "episodes about tolerance and acceptance reinforce deafness as a cultural trait, especially in storylines in which multiple characters speak in

American Sign Language."[58] Through examples from *Switched at Birth*, Foss reveals how the show presents Deaf characters as deficient and less capable than their hearing peers; in contrast, she also shows how the show can be multifaceted, covering storylines on essential topics that have nothing to do with deafness. The characters' use of ASL on *Switched at Birth* frequently is portrayed as "incomplete and simplistic, downplaying its grammatical complexity."[59] While considering the recent research on deaf characters, we must acknowledge that some of the *Switched at Birth* audience are viewing deaf characters on television for the first time.

The Switched at Birth *Audience*

Producers make decisions about how the audience is given access to a minority language, and they may include subtitles of foreign speech, translations of foreign speech through other characters, or allow the language to remain inaccessible except for those fluent in the language. While it is not unusual to hear another language spoken and read English subtitles, it is rare to exclude the second language from viewer access. Yet, frequently the filming techniques and editing used in the series removes the ASL from the audience's view. Viewers rely on a subtitled English interpretation for access.

As previously noted, new technology enables close-ups to be included with more frequency than in the past. In addition, changes in how viewers are accessing films and television along with budget restraints have made close-ups increasingly popular. Wide shots do not translate as well on smaller devices, such as computers or phones, where many viewers are watching movies and television with increased frequency; close-ups become easier for people to see. "Cinematographers also point to the rise of 'video assist,' a system where filmmakers shoot while watching monitors on set."[60] Giving preference to close-up shots inadvertently eliminates viewers' access to ASL since the language requires the viewer to see the use of hands, arms, and the torso. Further, "film historians argue that a tight shot on a digital camera can look more like actual film, a rich quality that many directors prefer."[61] At times the close-ups are for dramatic purposes, and "it's never been easier to create the perfect close-up with software that erases wrinkles and other imperfections pixel by pixel."[62] As Scott Higgins, film studies chair at Wesleyan University explains, "if you want to make a scene intense, one way to do it is to get in close."[63] Since close-ups are seen more frequently in movies and television, the audience not fluent in ASL may be unaware that language is being eliminated since they are reading English captions; in fact, since the characters are so skilled in lipreading, the audience may perceive this as clear communication that occurs between deaf people.

It is understandable that the producers of *Switched at Birth* are following industry trends by making choices to include close-ups; yet, these trends work against the accurate portrayals of Deaf characters. Storylines including Deaf cultural values must atone for the filming techniques that exclude one of the most valued aspects of the culture: ASL. Arguably, the series compensates by including other cultural values.

Even though American viewing habits increasingly exclude foreign films, Americans are becoming more comfortable with films and television shows incorporating English subtitles.[64] Scott Foundas explains, "American audiences seem more willing than ever to embrace cultural and linguistic diversity as a cinematic norm, and less accepting of the old Hollywood convention by which everyone in every corner of the world happens to speak perfect, American-accented English."[65] Including several Deaf characters who use ASL as their main form of communicating was somewhat risky for the network considering that adding this form of diversity requires that all viewers, Deaf and hearing, rely on English subtitles. With Deaf characters being central to the storyline, the audience must read subtitles for a good portion of each episode. Even for Deaf viewers and those fluent in ASL, one must read the subtitles because through close-ups and shifts in scenes, much of the signing is eliminated. Fortunately, "subtitles have become increasingly common at the mainstream multiplex in the past two decades, with no audience revolt in sight."[66]

If the Switch Had Never Happened...

Essentially, many of the storylines on *Switched at Birth* are written for a "hearing" audience to demonstrate the idea that an existence completely different from one's own would be equal in experience. While some episodes feature deafness as a plot device, many fit into Lerner's classifications of "stories about deaf/hearing relationships" and "a-normal-guy-or-gal-who-just-happens-to-be-deaf." Overall, appreciating one's identity as a Deaf person, or Deaf Pride, is a significant cultural value within the Deaf community.

The writers of *Switched at Birth* attempt to convince the audience that the characters we have grown to love are here because of the circumstances of the switch. The series debut entitled "This Is Not a Pipe"[67] gives viewers a few hints about what life would be like had the switch never happened. In a conversation between Emmett and Daphne, Emmett mentions the domino effect and how one small event has the power of influencing all events that follow. What if Daphne had grown up with the Kennish family instead of with a single mother? This points to Daphne's deafness being contingent on being raised in a poorer economic situation than with the well-off Kennishes.

When Emmett sees Bay for the first time, he asks Daphne who she is. Daphne's response is "Me. In another life." While this first episode does not go into any more detail about the hypothetical life where the switch had not happened, the series addresses this issue in other episodes throughout the series, explicitly in "Ecce Mono" and "Paradise Lost."[68]

The episode "Ecce Mono" begins with John Kennish and Regina Vasquez arguing; John has a heart attack and dreams that all of the children had been raised in the Kennish household. Without Regina's influence, the personalities of Daphne and Bay are quite different. Daphne does not know ASL and uses a cochlear implant, a surgically implanted electronic device that replaces the function of damaged inner ears by providing simulated sound signals to the brain. Some misunderstand the function of the cochlear implant, believing it enables a deafened individual to regain a normal level of hearing. It does not make a user hear again but it can help users recognize sounds that are around them, such as alarms and doorbells. While today cochlear implants are more generally accepted as another form of assisted listening devices, to some, they are perceived as a type of genocide within Deaf community. In the episode, Daphne with a cochlear implant appears callous towards her family members. Bay, attempting to compensate for being the adopted-into-the-family child, is a straight-A student. In one scene, Daphne drags Bay to a party. In this other reality, Emmett happens to be driving by the party on his motorcycle when he sees Bay leaving and walking alone. Even without Bay knowing ASL, he convinces her that she should not be walking alone at night and gives her a ride. At the end of the episode, they ride off together.

"Ecce Mono" shows how Bay and Emmett, regardless of the circumstances, are destined to be together. It also removes Regina's role of meeting Melody Bledsoe (Marlee Matlin) and her son Emmett who introduce them to ASL. In this reality, ASL is missing from the lives of Daphne and the Kennish family. The removal of Regina and ASL presents Daphne as a snobby, unfriendly character. Similar to how Lerner classifies ASL as saving the day, here ASL saves the family. At the very least, a Daphne who knows ASL is gentler and kinder than one who does not.

The episode concludes that if it were not for Regina, John Kennish would have had a heart attack alone and he would have died. Because she exists and calls for an ambulance, John opens his eyes and says, "Thank God you're here," showing how grateful he is for how all of their lives turned out. But we cannot thank Regina without recognizing Emmett and Melody's involvement in Daphne's formative years. Regina and Daphne know ASL because of the Bledsoe family. Again, it is the culturally Deaf characters played by Deaf actors who offer the depth in the storyline. Not only has ASL saved the family, but when unpacking the scene we must recognize that there was a collective effort by Deaf people to assist another member of their community.

Daphne dreams that the switch had never happened in the episode "Paradise Lost." The scene begins with Daphne (Leclerc) living her life as Bay. A typical morning with family leads her to walk by a mother and daughter conversing in ASL. This is Regina and "Daphne," who for this part of the episode is played by Vanessa Marano. As Bay (Leclerc) stares at the girl, Emmett rides up on his motorcycle and asks Daphne (Marano) what is wrong. Daphne (Marano) responds that this girl just keeps staring at her. Emmett suggests that the two of them give the girl something to stare at and they start kissing. Daphne's dream includes an alternative life where Daphne, who is in Bay's body, is now the deaf girl who still ends up with Emmett. Daphne (Leclerc) wakes up from her dream in a gasp.

This episode is much more about Daphne coming to terms with her feelings for Emmett. What I enjoy about the scene is the suggestion that Bay and Emmett, affectionately referred to as Bemmett by fans, are still together in this alternative reality.[69] Throughout the series, the two are connected through their love of art which transcends being Deaf or hearing. Later in the episode, after Daphne wakes up startled from her dream, Emmett attends Bay and Daphne's birthday party. He gives a *Deafenstein* movie poster to Daphne. *Deafenstein* is the fictional horror movie that is performed in ASL and which is mentioned in several episodes throughout the series. It appears as a cult movie similar to the actual movie *Deafula* (1975) by Peter Wechsberg, which was written, produced, and cast by Deaf people.[70] For Bay's gift, Emmett takes her to a billboard he has tagged. Emmett and Daphne share a love for Deaf cultural experiences, while he and Bay appreciate art. Viewers have an advantage since the prior scene includes his mother, played by Marlee Matlin, giving him the statistics that 85 percent of Deaf and hearing relationships do not work out. She follows up by asking, "Do you think she'll ever get to know you like a deaf girl would?" Emmett responds, "There's only one way to find out." Emmett shows pride in his culture but does not exclude other interests or isolate himself. Lerner might classify this scene as one portraying deaf/hearing relationships and the "normal" deaf guy, yet to viewers, Emmett is the guy many aspire to date. This scene mirrors reality, as Matlin has shared, "I knew from the start that there was something unique and groundbreaking about the show," citing how ASL and subtitles had been incorporated into the script "in a manner that [she] had only dreamed should happen in TV."[71] Matlin describes the show as "a game changer for the community of deaf actors in Hollywood, as well as viewers eager for diversity."[72] Having Emmett, a Deaf character played by a Deaf actor, nonchalantly become the heartthrob of the show is exactly what the Deaf actors of the past have worked towards. This scene is the passing of the torch from one Deaf actor to another. Matlin as Melody cautions Berdy as Emmett about the unfortunate realities of the world; Berdy decides to take a chance against the odds.

Both of these episodes remind viewers that despite the circumstances of the switch, the characters are exactly how they are supposed to be. The role of deafness and the inclusion of ASL in the episodes is part of the diversity of the characters. The charm of the episodes comes how likeable, believable, and authentic the characters are portrayed.

After five years and over 100 episodes, viewers have followed the *Switched at Birth* storyline of two girls and their families grappling with making connections, working together, and viewing themselves as a family. The characters have learned about each other, and we, the viewers, have as well. The audience also sees what a family should do: love their children for who they are. In this case, that means that they all learn ASL as a family. This is not a reality for many Deaf people, so even if the parental characters on the show did not become proficient communicators in ASL, they at least tried. After all, it has only been five years. It takes much longer to become fluent in a language; we must assume that these characters will continue to learn from one another.

In the Merriam-Webster dictionary definition of the word "deaf," it is defined as "lacking or deficient in the sense of hearing" and "unwilling to hear or listen."[73] Even conversational language concerning deafness is embedded with pejorative uses, and I cannot imagine any young viewer of the television show approving of these definitions. The Deaf characters in *Switched at Birth* have been innovative, daring, inquisitive, caring, and even ordinary when viewers are allowed to view them as everyday teens. The network spent a great deal of effort incorporating Deaf cultural values in order to engage viewers. These cultural values, including the appreciation of ASL, Deaf schools, and one's identity, are shown to be important aspects of the Deaf characters. While the episodes have been entertaining, they have also been educational in that viewers learned some Deaf history, and how best to interact with Deaf individuals.

Switched at Birth demonstrates how including cultural values, and how working with Deaf actors, film makers, and writers can create a more realistic depiction of deafness and Deaf people. Television and films do not always portray the most culturally accurate characters. Unfortunately, even with discussions of valuing ASL, the show has not been consistent in showing the language in use. Following filming trends which emphasize close-up shots, the film crew has inadvertently removed ASL from the frame. Yet, *Switched at Birth* has shown ASL more frequently than any other mainstream series of its time, thereby moving in the right direction.

The inclusion of more Deaf writers, directors, and producers would continue to increase positive depictions of culturally Deaf characters. Through a value of collectivism, Deaf actors will continue to work together to fight for roles. *Switched at Birth* is another stepping stone for future shows. The actors, including guest stars like Nyle DiMarco, are paving the way for future series incorporating even more accurate portrayals of Deaf characters.

Notes

1. Gail Pennington, "TCA: 'Switched at Birth' Goes Deeper into Deaf Culture," *St. Louis Post-Dispatch*, January 12, 2013, http://www.stltoday.com/entertainment/television/gail-pennington/tca-switched-at-birth-goes-deeper-into-deaf-culture/article_d91c3105-de2a-5167-a313-f88074d19afa.html.

2. In accordance with literature of the field, I refer to "Deaf" as representing individuals who identify in a linguistic, cultural minority group. The term "deaf" is used as a more generic term given to individuals with some degree of hearing loss. In other articles, "deaf" has been used pejoratively or in connection to a pathological view by educators, doctors, counselors, and members of the larger society who believe one without the sense of hearing is inferior or lacking. I do not believe or wish to imply that at all.

3. Eugene Bergman, "Literature, Fictional Characters," in *Gallaudet Encyclopedia of Deaf People & Deafness*, ed. J.V. Van Cleve (Gallaudet College, Washington, D.C.: McGraw Hill, 1987), 172–176.

4. John S. Schuchman, *Hollywood Speaks: Deafness and the Film Entertainment Industry* (Urbana: University of Illinois Press, 1999), 65.

5. Ellen Gamerman, "Hollywood's Extreme Close-Up: New Technology has Helped Make Close-Ups a Common Tool for Filmmakers, Changing How Movies Look, Actors Perform and Stories Unfold," *Wall Street Journal*, January 21, 2017, http://www.wsj.com/articles/the-close-up-close-up-1485000003.

6. Along with valuing American Sign Language, members of the cultural Deaf community share similar values. For example, while the dominant culture in the United States includes *individualism*, members of the Deaf community view themselves as an interconnected group focusing on *collectivism*. Upon meeting, many cultural Deaf individuals attempt to identify the Deaf friends they both have in common. As another example, unlike members of the dominant culture who may find it rude to point out a person's physical appearance, this is noted within Deaf culture as it directly connects to a visual language. One last example comparing the dominant culture with Deaf culture is that ASL requires one to look at the person; broken eye contact or looking away is considered rude, whereas members of the dominant culture find that looking at someone too long, or staring, is considered rude.

7. Melanie E.S. Kohnen, "Cultural Diversity as Brand Management in Cable Television," *Media Industries* 2, no. 2 (2015): 91.

8. Lizzy Weiss, "Uprising," *Switched at Birth*, season 2, episode 9, ABC Family, March 4, 2013.

9. The episode begins and ends with spoken dialogue.

10. Carol Padden and Tom Humphries, *Deaf in America: Voices from a Culture* (Cambridge: Harvard University Press, 1988).

11. Kristen Harmon, "Addressing Deafness: From Hearing Loss to Deaf Gain," *Profession* 2010, no. 1 (2010): 124–130.

12. Here in America, Deaf community members use ASL. It is not, however, a universal language. Deaf communities across the world have their own distinct forms of signed language.

13. Trent Batson, "The Deaf Person in Fiction: From Sainthood to Rorschach Blot," *Interracial Books for Children Bulletin* 11, no. 1, 2 (1980): 18.

14. Carol Erting, "Cultural Conflict in a School for Deaf Children," *Anthropology and Education Quarterly* 16, no. 3 (1985): 225–243.

15. Doug Alker, "Misconceptions of Deaf Culture in the Media and the Arts," in *The Deaf Way: Perspectives from the International Conference on Deaf Culture*, ed. Carol J. Erting (Washington, D.C.: Gallaudet University Press, 1994), 723.

16. Melanie E. S. Kohnen, "Cultural Diversity as Brand Management in Cable Television," *Media Industries* 2, no. 2 (2015): 88.

17. *Ibid.*, 92.

18. *Ibid.*

19. *Ibid.*, 94.

20. *Ibid.*

21. Examples include Daphne applying for the Latina Scholarship and Natalie being comfortable as a Deaf Lesbian.

22. Chuck Barney, "Chuck Barney: 'Switched at Birth' Another Winner for ABC Family," *San Jose Mercury News*, June 29, 2011.

23. Schuchman, *Hollywood Speaks*, 3.

24. *Ibid.*, 5.

25. Ann Silver, "How Does Hollywood See Us, and How Do We See Hollywood?" in *The Deaf Way: Perspectives from the International Conference on Deaf Culture*, ed. Carol J. Erting (Washington, D.C.: Gallaudet University Press, 1994), 735.

26. Schuchman, *Hollywood Speaks*, 7.

27. Patti Durr, "Deaf Cinema," in *The Sage Deaf Studies Encyclopedia*, ed. Genie Gertz and Patrick Boudreault (Thousand Oaks, CA: Sage, 2016): 157–158.

28. Phyllis Frelich, "Hollywood through Deaf Eyes: A Panel Discussion," in *The Deaf Way: Perspectives from the International Conference on Deaf Culture*, ed. Carol J. Erting (Washington, D.C.: Gallaudet University Press, 1994), 737.

29. *Ibid.*

30. Silver, "How Does Hollywood See Us, and How Do We See Hollywood?" 732.

31. *Ibid.*, 733.

32. S. E. Smith, "Why Is Hollywood Still Stubbornly Casting Nondisabled Actors in Disabled Roles?" XOJane.com, February 4, 2015, http://www.xojane.com/issues/cake-still-alice-cripface-oscars.

33. Silver, "How Does Hollywood See Us, and How Do We See Hollywood?" 734.

34. Linda Buchwald, "Deaf Talent, Seen and Heard," *American Theatre*, October 20, 2015, http://www.americantheatre.org/2015/10/20/deaf-talent-seen-and-heard/.

35. Jules Dameron's Twitter page, https://twitter.com/julesdameron.

36. Deaf Talent's Facebook page, www.facebook.com/deaftalentnow.

37. Jillian Leff interviews Katie Leclerc, *Switched at Birth Star Katie Leclerc Talks Ménière's Disease*, YouTube, 35:25, March 6, 2014, https://www.youtube.com/watch?v=skABO0z-fxc&feature=youtu.be.

38. Don Grushkin, "What Do Deaf People Think of the Show *Switched at Birth*? Discounting the Soap Opera-Like Storylines, Is It a Realistic Depiction of Deaf Life?" *Quora: Deafness (physiological condition)*, February 14, 2015, https://www.quora.com/What-do-deaf-people-think-of-the-show-Switched-at-Birth.

39. Martha Anger's Facebook Live video, March 30, 2017, https://www.facebook.com/martha.s.anger/videos/10102185839332495/.

40. Rob Nielson, "Create Responsible, Accurate, and Family-Oriented TV Programming," *Change.org*, 2012, https://www.change.org/p/abc-family-and-the-switched-at-birth-series-create-responsible-accurate-and-family-oriented-tv-programming.

41. Don Grushkin, "What Do Deaf People Think of the Show Switched at Birth? Discounting the Soap Opera-Like Storylines, Is It a Realistic Depiction of Deaf Life?"

42. Smith, "Why Is Hollywood Still Stubbornly Casting Nondisabled Actors in Disabled Roles?"

43. *Ibid.*

44. Jennifer Rayman, "The Politics and Practice of Voice: Representing American Sign Language on the Screen in Two Recent Television Crime Dramas," *M/C Journal* 13, no. 3 (June 30, 2010), http://journal.media-culture.org.au/index.php/mcjournal/article/view/273.

45. *Ibid.*

46. Miriam Nathan Lerner, "Narrative Function of Deafness and Deaf Characters in Film," *M/C Journal* 13, no. 3 (June 30, 2010), http://journal.media-culture.org.au/index.php/mcjournal/article/view/260.

47. Seon-Kyoung An, Llewyn Elise Paine, Jamie Nichole McNiel, Amy Rask, Jourdan Taylor Holder, and Duane Varan, "Prominent Messages in Television Drama *Switched at*

Birth Promote Attitude Change Toward Deafness," *Mass Communication and Society* 17 (2014): 195–216.

48. *Ibid.*, 195.

49. *Ibid.*, 211.

50. Schuchman, *Hollywood Speaks*, 6.

51. *Ibid.*, 211.

52. Katherine A. Foss, "(De)stigmatizing the Silent Epidemic: Representations of Hearing Loss in Entertainment Television." *Health Communication* 29, no. 9 (2014): 888–900.

53. *Ibid.*, 891.

54. "Avenged Filmmaker Defends Decision Not to Use Deaf Actress as His Lead," *WENN in Movies*, March 11, 2015, http://www.contactmusic.com/news/avenged-filmmaker-defends-decision-not-to-use-deaf-actress-as-his-lead_4627967.

55. Thomsen Young, "Why #deaftalent Means Something," *Silent Grapevine*, February 1, 2015, http://silentgrapevine.com/2015/02/why-deaftalent-means-something.html.

56. The Deaf character is played by actress Kate Siegel, who also helped write the script. This was not an acceptable decision by members of the Deaf community. See "Netflix's 'Hush' Robs #DeafTalent," *The Daily Moth*, April 8, 2016, https://www.youtube.com/watch?v=NUr Wolo Qpf0.

57. Katherine A. Foss, "Constructing Hearing Loss or 'Deaf Gain?' Voice, Agency, and Identity in Television's Representations of d/Deafness," *Critical Studies in Media Communication* 31, no. 5 (2014): 426.

58. *Ibid.*

59. *Ibid.*, 440.

60. Ellen Gamerman, "Hollywood's Extreme Close-Up."

61. *Ibid.*

62. *Ibid.*

63. *Ibid.*

64. Anthony Kaufman, "The Lonely Subtitle: Here's Why U.S. Audiences Are Abandoning Foreign-Language Films," *Indie Wire*, May 6, 2014, http://www.indiewire.com/2014/05/the-lonely-subtitle-heres-why-u-s-audiences-are-abandoning-foreign-language-films-27051; Scott Foundas, "Why U.S. Audiences Are More Comfortable with Subtitles Than Ever," *Variety*, April 22, 2014, http://variety.com/2014/film/columns/why-u-s-audiences-are-more-comfortable-with-subtitles-than-ever-1201160162/.

65. Foundas, "Why U.S. Audiences Are More Comfortable with Subtitles Than Ever."

66. *Ibid.*

67. "Pilot," Switched at Birth, season 1, episode 1, ABC Family, June 6, 2011.

68. "Ecce Mono," *Switched at Birth*, season 2, episode 15, ABC Family, July 8, 2013; "Paradise Lost," *Switched at Birth,* season 1, episode 9, ABC Family, August 1, 2011.

69. Fans add the characters names Bay and Emmett together as a power couple, referring to them as Bemmett.

70. This is an incredible nod to DeafTalent since *Deafula* (1975) was created by and for Deaf people.

71. David M. Perry, "How to Break Ground for Deaf Actors in Hollywood," *Pacific Standard,* April 11, 2017, https://psmag.com/how-to-break-ground-for-deaf-actors-in-hollywood-9e13928e795c.

72. *Ibid.*

73. "Deaf," *Merriam Webster*, https://www.merriam-webster.com/dictionary/deaf.

Models and Misbehavior

ABC Family's Portrayals
of Sexual Health Topics

MALYNNDA A. JOHNSON
and KATHLEEN M. TURNER

Lessons about relationships and appropriate behaviors have been passed to audiences through the use of narratives since the early Greeks. In modern times, these stories are often presented in the form of television shows, internet videos, and movies. From lessons about how to ask someone out on a date to when you are ready to have sex, today's entertainment industry presents interpersonal lessons for viewers, especially in terms of romantic and sexual relationships. Since previous research suggests that young adults, in particular, spend close to half of their day engaged with some form of media, young audiences are perhaps among the most affected by lessons from mediated narratives.[1]

Narratives, or stories real or imagined, provide glimpses into personal experience. Such stories are essential to the human experience, providing a common structure composed of characters, conflict, lesson and a resolution.[2] Lured in and engaged by the drama of a narrative, audience members are taught lessons while sharing a common experience. Evolving from oral histories, to print mediums, to film, television, and now internet video services, mediated narratives now provide a greater reach in audience. Today, countless sources provide a wide range of narratives 24 hours a day. As a result, networks are forced to find ways to produce content that will captivate their audiences. To compete for viewership, entire television networks, including ABC Family/Freeform and the CW are tailoring their programing toward young audiences.

According to Glen Sparks, an average adolescent spends close to 8.5

hours a day with television or internet.[3] As their age increases, so does the number of hours spent with these mediums; college students, for instance, increase their average to 12 to 13 hours a day.[4] While many forms of media, including print, radio, or the internet, contain messages focusing on sex, in terms of the quantity of sexually implied content, research suggests that the messages mediated through primetime television are relatively more consistent than these other mediums.[5] When it comes to mediated stories of sex and relationships, the percentage of TV programs containing sexual content, showing or implying sexual behavior, has significantly risen, from 56 percent in the 1997–98 television season to 70 percent of shows in 2004–05.[6] In the 2004–05 TV season, slightly more than 1 in 10 programs (11 percent) included some portrayal of sexual intercourse.[7] However, due to network regulations and Federal Communications Commission (FCC) regulations, most programs rarely show explicit portrayals of sexual intercourse. More recently, nearly 75 percent of primetime shows aired in 2008–2009 contained some form or mention of sexual content, with only 14 percent of the sexual incidents mentioning any risk or consequence for such activities.[8] However, today's young viewers are witnessing some of the risks associated with sexual activity. Shows such as *16 and Pregnant* (MTV, 2009–present), *Teen Mom* (MTV, 2009–present), and *The Secret Life of an American Teenager* (2008–2013) have specifically directed attention to the consequences of sex, often focusing on the reality of unplanned pregnancy. However, seldom have shows included sexually transmitted illness as a theme of sexual consequence in television.

Drawing connections between the consumption of media and the engagement of sexual activity among young people today seems to have redefined what it means to engage with media and each other. Therefore, phrases about television viewing habits (i.e., "binge watching") and sexual activity (i.e., "hooking up") are prevalent among today's young adults. The ubiquity of colloquial phrases surrounding both television and sexual activity gave rise to the recent phrase "Netflix and chill," which is an invitation for sexual activity under the guise of watching Netflix, which directly links television viewing and sexual activity. Knowing that numerous lessons about sex abound in popular media today, we sought to explore the mediated lessons being presented to tweens and teens.

Throughout their programming, ABC Family/Freeform shows portray teens and young adults struggling with their decisions to become sexually active. In some shows, young characters take purity pledges and struggle with losing their virginity. Thus, many of the early shows seemed to align well with ABC Family's original Christian Broadcasting Network framework (e.g., *Make It or Break It,* 2009–2012, and *The Secret Life of the American Teenager*). However, over the years, these shows' topics and characters have matured, resulting in less of a focus on "punishing" sexual promiscuity to arguably promoting sexual activity, even between students and teachers (e.g., *Pretty Little Liars,* 2010–2017). Critical rhetoric reveals several important themes as we consider how these narratives punish or promote sexual activity for teenage characters and potentially teach young viewers about the dangers of sex, while extolling the virtues of virginity, especially in the case of the youthful women in the shows. Overall, this essay analyzes the changing trends of how ABC Family/Freeform has presented narratives of young adults in relation to topics of sexual health.

Television's Impact on Behavior

Victor Strasburger and colleagues argue, "television … now escorts children across the globe even before they have permission to cross the street."[9] Television is also thought to represent an extension of one's senses while offering an experience, a window, into other worlds.[10] Milly Buonanno states that while creating a sense of place, television provides viewers both symbolic and imaginative mobility.[11] With this new mobility, viewers are given the opportunity to encounter places and situations they have never experienced before. While the ability for viewers to explore these new worlds from the comfort of their own homes offers excitement and possibility, what remains unknown is what viewers take away from these mediated voyages. Following the notions of George Gerbner's cultivation theory, viewers do more than simply watch a program—they absorb it.[12] Doing so, viewers become more than just viewers; they become involved with the medium, caring about what happens and even interacting with the characters by talking to the show.[13]

Numerous theories have been developed to explain how television shapes perceptions and behaviors. Therefore, to make the argument that perceptions of sexual behavior have been shaped by mediated narratives, it is important to understand how television educates audiences regardless of the intentions to do so. Central to many media theories, and one of the most widely investigated concepts, is involvement.[14] William J. Brown argues that involvement, when broadly defined, is the degree of a person's psychological response to a mediated message.[15] Thus, involvement should not be seen as a static or fixed idea. Rather, it is a dynamic process through which a viewer can become motivated both during and after media consumption.[16] While the majority of involvement research has focused on marketing and advertising studies, persuasion scholars have studied the various levels of a viewer's involvement with mediated messages and the connections between this involvement and motivation to alter one's behavior.[17] For example, if viewers watch a character of a similar age, gender, and set of interests, they will be more likely to connect with the character. As a result of this new sense of connection, the viewer may begin to form parasocial, or one-sided, relationships with the character. Given the viewer's perception of similarity to and connection with the character, a viewer's involvement with a show or character may have a more powerful influence on their behavior than initially suspected.[18] When observing poor choices and negative behaviors, audiences are not only more likely to watch, but studies have found that these messages tend to stick with the viewer longer.[19]

Research has also found that viewers take more from shows they already like or watch on a regular basis.[20] Therefore, if a strong connection to a character results in the viewer being less likely to challenge the messages provided, the level of involvement may also be inferred to be high, and the questioning of content will probably be low.[21] The key to understanding involvement within these social models is to understand how the viewers are connecting with the show, as well as how they resonate with the characters.

Consistently, ABC Family has provided diverse characters with whom viewers can identify including a range of ages, races, economic backgrounds, even sexual orientations. Beyond the characters themselves, viewers are also able to observe situations they themselves might experience. From fighting with parents, peers, and siblings, to decisions about when to become sexually active, viewers come face to face with the realities of growing up in a time where sex and drama equals ratings.

Framing Sexual Health Within Television Programing

Rhetorical scholar Kenneth Burke explains in his text *Attitudes Toward History* that presenting lessons within dramatic narratives occurs through the use of either tragic or comic frames.[22] Within the tragic frame, "evil" or wrongdoing is placed on the shoulders of a scapegoat. A person is made an example of, or used as a tool for evil doing. Often the scapegoat is a secondary character who engaged in risky behavior beyond the knowledge of the central character. For example, to teach a lesson about underage drinking, the character hosting the party and providing the access to alcohol will be to blame. As a result, it will likely be his or her house that is destroyed, or another unfortunate action will occur leading to police or parents becoming involved. Ultimately, the tragic frame seeks to blame and punish, therefore restoring the social order.

Alternatively, the goal of the comic frame is not to mortify the character. Instead, the comic frame seeks to "teach the fool—and vicariously the audience—about error so that it may be corrected rather than punished."[23] Doing so allows for the character, and viewer, to learn an important lesson while still remaining part of the larger community. By implementing the comic fame, shows can present lessons about the potential consequences of promiscuous sex or drug use while also maintaining a sense of shared community and allowing the parasocial relationships to remain intact. As a result, the viewer is taught that the offender can be reformed. The character will likely feel some level of guilt for their behavior; however, their guilt is often assuaged upon discussing events and actions with other characters. Once "reformed," the character is able come back to the community, thus maintaining the social order.

Applying Burke's ideas to the narratives surrounding sexual behaviors of ABC Family/ Freeform shows provides insight into the dilemmas and paradoxes associated with including sexual health topics and behaviors in television. When shows are successful in merging entertainment and education about sexual health behaviors, observers will be able to see the implications of their actions. The popularity of ABC Family/Freeform's programming clearly demonstrates that young viewers are engaged in the shows. The question remaining is, what are viewers potentially learning from these shows?

Methods

In order to unpack the potential lessons presented in ABC Family/ Freeform shows, a close reading of a variety of shows was conducted. Spanning a nine-year timeline, five shows were selected based on theme and ratings. Shows ranged from some of the most popular ABC Family shows to some of the more recent Freeform programs, and two shows which were on the air during the rebranding of the channel to Freeform in 2016 are included. Given that the first season of a television show sets the tone for the season, we focus this study on season one for the following ABC Family/Freeform shows: *Greek* (2007–2011), *The Secret Life of the American Teenager* (2008–2013), *Pretty Little Liars* (2010–2017), *The Fosters* (2013–2018), and *Recovery Road* (2016). It should be noted that *Recovery Road* began shortly after the Freeform rebranding, and was cancelled after its initial season.

Analysis of ABC Family/Freeform Shows

In 2007, one year into the life of ABC Family, the network launched the comedy/drama *Greek*. The show centers around Casey Cartwright (Spencer Grammer) and her younger brother, Rusty (Jacob Zachar), both of whom are enrolled at the fictional Cyprus-Rhodes University (CRU), located in Ohio.

Depicting fraternities, sororities, and the Greek community as a whole in the most stereotypical ways possible, the show focuses the majority of its themes on drinking, partying, relationships, and when there is time, going to class.

Greek also places heavy emphasis on sex. Used as a tool for manipulation, status, and something to do for fun, the show offers multiple situations supporting the notion of a college "hookup" culture, or physical relationships with little to no expectation of corresponding emotional relationships. Beginning with the pilot episode, Casey's boyfriend, Evan (Jake McDorman), is caught having sex with Rebecca (Dilshad Vadsaria) by Rusty. Upon finding out about the indiscretion, Casey is told by her sorority's "big sister," Frannie (Tiffany Dupont), that she can't break up with Evan simply because he slept with someone else. Explaining further, Frannie more than suggests that it is the sexual relationship with Evan that will help Casey become president of their sorority. Seeking status within the sorority, Casey chooses to stay with Evan. The use of sex as manipulation is seen again in the episode "The Rusty Nail," when Frannie learns that Casey has not had sex with Evan since he cheated. Frannie again explains to Casey that she has to start having sex with Evan again "for the good of the sorority."[24] Leveraging sex to gain status in the sorority, as well as pressuring Casey to remain in an unhealthy relationship for the sake of the sorority, sets up dangerous and dysfunctional relationship tropes.

The use of sex as a tool for manipulation becomes an even more complicated issue when used to seduce unsuspecting administrators. When the Kappa Tau fraternity house is in need of a special permit for a party, the notion of using sex as a means of achieving a goal is extended. In episode five, "Liquid Courage," Cappie (Scott Michael Foster), Rusty's "big brother," is tasked with the seemingly impossible job of obtaining a noise permit required for an upcoming party.[25] Relying on his trusty "cougar sack," a toolkit for attracting women of an older age, Cappie sets off with Old Spice Ultra cologne and his Madras jacket, ready to win the affection of permit gatekeeper Gladys, an older woman with gray hair and a very "unique" sense of style, including bright blue eye shadow and a tawdry-looking sweater covered with large colorful sequins. Turning on the charm and smooth talking his way into her good graces, Cappie later returns to the house with the noise permit. As Cappie enters the house, his posture and downturned face implies a feeling of disgust as he quickly stops at the bar to take a shot. Although it is not clear what actually took place between Cappie and Gladys, given he was able to return to the house with permit in hand, the implication is that the plan of seducing the administrator in some fashion worked to gain the noise permit.

Wrapped into the show's portrayal of the Greek system are numerous gender stereotypes, especially surrounding the double standard with regards to sexual promiscuity. Women, especially sorority members, are frequently judged on their bodies and clothing by men and other women. Fraternity members are consistently shown cheering on their brothers as women walk out in the morning from having sex with their fraternity brothers, while women are quickly labeled as sluts when they engage in sex outside the bounds of a relationship.

With a heavy emphasis on sexual activity, it should be of little surprise to observe a negative framing of virginity. For the cast of *Greek*, virginity is something that needs to be overcome in order to assimilate into the hookup culture of college. In episode three, "The Rusty Nail," Rusty struggles to find a date for his first Greek life function with his fraternity. While this is not a concern for his brothers, Rusty feels an added layer of pressure given he has never been on a date.[26] In the episode, Rusty refers to his virginity as a "burden," making him feel like "a big awkward loser." Virginity is portrayed by Rusty and his fraternity brothers as an embarrassment that should end. As a rite of passage, Cappie diagnoses Rusty with an acute case of "virginitis," claiming that if "left untreated," Rusty's "30s will not be pretty." Cappie then claims Rusty needs to have sex so that he can stop thinking of women as "magical creatures, like a unicorn with breasts." Cappie quickly sets Rusty up with Lisa Lawson (Arielle Vandenberg), "the virgin whisperer." Rusty, however, leaves Lisa after she stops their foreplay to take a call from her mother. Offering Lisa as a means of curing Rusty's "virginitis" not only degrades Lisa to a nothing more than sexual object, but it also devalues the idea of wanting to wait for a romantic relationship before having sex.

When virginity is discussed with any semblance of religious value, it is only as a means of mocking those who might place value on their sexual purity. For instance, Dale (Clark Duke), Rusty's roommate, is a devout Baptist. When Dale learns Rusty has never been on a date, he suggests that Rusty join the Purity Pledge program and gives him a brochure and a sales pitch, which he ends by saying, "Just remember, your virginity is a sacred gift from God."[27] Later, Dale tries to get Rusty to join his purity pledge brothers, telling Rusty that Purity Pledge is "just like a fraternity, but for God." Dale shows Rusty the ring he wears as part of the Purity Pledge group, which is a large square ring with a giant P on it. This ring features prominently in shots of Dale throughout the rest of the first season. He talks about their rings and says, "Purity Pledge activate," and the guys put their purity pledge rings together as they all say "Amen."[28] Thus, the episode's allusion to Hanna-Barbera's Wonder Twins on *Super Friends* serves as comic relief, making fun of spirituality and purity pledges, using a cartoon superhero reference.

The jokes at Dale's expense continue with the invention of the "Dale Tracker." In the episode titled "Depth Perception," Rusty and his girlfriend, Jen K. (Jessica Rose), have had sex a few times, but they are having a hard time finding alone time in Rusty's room since his roommate Dale is always around.[29] Rusty installs a GPS device on Dale's cell phone and creates a "Dale Tracker" application on his computer so that he and Jen K. can be alerted anytime Dale is coming back to the room. Inevitably, Dale finds out about the "Dale Tracker," leading to a scene where he hides in the closet and jumps out as soon as Jen and Rusty are about to have sex.

Although the majority of sexual encounters on *Greek* fit the heteronormative storyline, the show does provide a few homosexual narratives. Not only does the show imply homosexual sexual activity, but it also provides an important discussion about accepting who you are and the coming out process. Calvin (Paul James), one of Rusty's closest friends and fellow freshman, is introduced in the show as he begins pledging a fraternity. Sneaking around to meet up with Heath (Zack Lively), both men initially defend their sexuality, stating, "I'm not gay; we were just drunk." However, by episode four, "Picking Teams," Calvin is shown having a frank discussion with his father that touches on his sexuality and the fact that his dad made him join a fraternity in return for paying for school.[30]

Calvin's dad talks to Calvin about the deep and meaningful relationships he built in the fraternity, but Calvin quickly points out that as a homosexual he will not be able to engage in some of the rites his father mentions, like getting married. Persisting, Calvin's dad asks if Calvin has come out to anyone. When he says no, his father responds, "Son, you don't have anything to be ashamed of." Calvin then explains: "I'm not ashamed and I'm great about being gay. It's just how everyone else reacts that sucks. When I came out in high school, I went from being Calvin Owens to that gay hockey player guy. And I'm more than that, and this is my chance to start fresh on my own terms. I'll tell who I want when I want, but to be honest, Dad, it's just not the most interesting thing about me."[31] For the viewers, Calvin and Heath represent some of the realities faced by homosexual youth during the coming out process.

One of the most common struggles young people face in coming out is the fear of how others respond to the news. When Calvin comes out to his friend, Rusty admits to struggling to find the right words of support that don't sound condescending or like he is granting Calvin permission for his sexual orientation, but ultimately, Rusty finds the words, stating, "I mean we're friends; it's not like you need my permission to be gay."[32] Rusty confirms their friendship and lets Calvin know that his sexuality is just one facet of him. Later in the season, in the episode "Black, White, and Read All Over," Calvin comes out again, this time to his friend, Ashleigh (Amber Stevens West), when she makes a pass at him. Feeling hurt and embarrassed, Ashleigh becomes visibly upset and leaves before Calvin can discuss the whole situation with her.[33] Later in the episode, Ashleigh tells Casey about what happened, and Casey tells her, "maybe it's not about you." Continuing, Casey explains, "This is the Greek system. We're not always a beacon of tolerance. Calvin's probably worried about how people will react. I bet his biggest fear is that his friends will turn their backs on him." Casey highlights the kind of homophobia that permeates the Greek system,[34] and when Ashleigh takes Calvin an I'm-sorry present, his fraternity brothers immediately begin to tease her about being his girlfriend. When she claims she's not, one of the brothers sarcastically exclaims, "Oh, right, because Calvin is gay." Not realizing he's joking, Ashleigh says, "Oh, so you know too." Realizing her mistake too late, she rushes to apologize to Calvin. Returning to the fraternity, his brothers become very quiet, and many turn away from him as he says, "Hey guys. I guess you heard I'm gay.... Does anyone want to talk about it?"[35] As all the guys turn away and ignore him, Calvin removes his pledge pin and leaves the house. The message in Calvin's storyline demonstrates that while being sexually promiscuous is easily accepted within the Greek system, being open about one's non-heterosexual sexual orientation is not. Calvin loses his brothers and his house by virtue of his sexual orientation.

Given that ABC Family had previously been a Christian network, viewers might expect a holdover of themes and content aligned with Christian values. It is then surprising that, in the show *Greek,* choosing to abstain from sex is frequently mocked, and sex is shown as having few consequences. Contrary to common Christian behaviors (i.e., helping others, treating people with respect, reserving the body as sacred space), the viewer learns lessons about using the body (sex) for manipulation, shaming, and achieving personal satisfaction. Throughout the first season of *Greek*, it is clear that for members of the Greek system, sex is nothing more than a tool. First, sex can be used for status and power. Those who have not yet had sex are placed lower in the hierarchy from the outset. As a means of keeping the virgins out of positions of power, comic frames are used to maintain social order. The virgins are given opportunities to redeem themselves by having sex and joining in on what that society values. Those who refuse become the scapegoats and are targets for shame and harassment. Once characters have obtained sexual status, they then learn how to wield their power to get what they want. Thus, sex becomes a tool for manipulation. Characters are seen seducing and using their partners, or remaining in dysfunctional relationships, simply to achieve a goal. What is potentially the most concerning for a viewer is the pressure from the Greek "brothers" and "sisters" to engage in risky, and demeaning, behaviors; often for the perceived good of the sorority/fraternity. The pressure placed upon young people by society and peers alone is damaging, but when your "family" tells you what you should do and how you should act, the persuasion become even more dangerous.

Greek's depiction of college, and specifically the Greek life culture, is also problematic for viewers' perceptions of college life in general. Viewers of this show, and the ABC Family channel in general, are often middle and high school age; thus, these images begin to shape the expectations of college life. Overall, *Greek* rarely focused on the potential positive aspects of the college experience or the Greek system, such as forming friendships or gaining leadership experience. Instead, producers choose to focus on sex, partying, and manipulation.

Even before viewers get to college, ABC Family presents unhealthy and unrealistic relationships in the high school setting. Take, for example, *The Secret Life of the American Teenager*, a show centered entirely on the consequences of teenage sex in high school. Unlike the lighter, more comic frames found in *Greek*, the lessons taught in *Secret Life* tended to fall more on the tragic end of the spectrum. Grounding the shows' themes and lessons deep in Christian rhetoric, *The Secret Life of the American Teenager* appears in stark contrast to *Greek*.

The Secret Life of the American Teenager follows 15-year-old Amy Juergens (Shailene Woodley) as she learns that after having sex once while at band camp, she is now pregnant. Amy struggles with telling her friends and family, deciding what to do with the child, and trying to have a normal teenage relationship with her boyfriend, Ben (Ken Baumann). By comparison, other stereotypical young women are included, as if to show the basic "types" available to teenage girls. Amy's neighbor, Adrian (Francia Raisa), is shown as sexually promiscuous, while another classmate Grace (Megan Park), serves as a foil in the show playing the good Christian girl. As the daughter of a pastor, Grace takes an abstinence pledge in the first episode of the show and is observed openly proselytizing to everyone she can.

Overt religious discussions of sex throughout the show function as a redemption narrative. Consistently, characters are rewarded for good behavior and punished for bad behavior. Episode one, "Falling in Love," opens with Amy coming home and sneaking into the bathroom to pull a pregnancy test out of her French horn.[36] After she pulls out the test, the show cuts to her mother heating up dinner for her, before cutting back to a shot of Amy looking in the mirror with tears in her eyes. Initiating a tragic frame, Amy is ultimately punished for her indiscretion. The next day at school, her two friends race up to tell her about the new hot guidance counselor, and as she stares at them, they challenge her, questioning if she has anything better to talk about. Amy says simply, "I had sex." Her friends are incredulous and ask for more details. Finally telling them the tale, she explains, "It was not that great…. I'm not even sure it was sex, okay guys…. I don't know. I didn't exactly realize what was happening until like after two seconds, and then it was just over. And it wasn't fun, and definitely not like what you see in the movies, all romantic and stuff."[37] Amy makes the sex everyone else has been discussing as wonderful and exciting sound like a terrible thing. Her pun-ishment for behaving badly thus is not only a pregnancy but also the fact that the event leading up to it was one short, unpleasant night of coitus. Thus, from the beginning of the show the viewer is presented with strong Christian overtones. Given the channel's history and the partnership the show had with Christian abstinence-only groups, such as the Candies Foundation, such themes are not unexpected.

Throughout the first season, much of the discussion centers on purity, abstinence, and the importance of vows of chastity due to wanting to uphold Christian values. For example, in the episode "Falling in Love," while boys may drool over Grace, it is known around the school that she will not have sex because she's prioritizing her Christian faith and waiting until marriage.[38] When Grace is sitting down to have lunch with her boyfriend Jack (Greg Fin-ley), he asks, "What's that ring? I never noticed you had a ring like that." Grace explains, "It's a promise ring. My parents gave it to me when I promised I wouldn't have sex until I get married." Jack states that he knows what a promise ring is, but he didn't know she had one. Grace says, "Last night my parents and I had a long talk about you and me." Jack seems a little upset and says he thought her parents trusted her and that he's a Christian who is "just as committed to abstinence" as Grace is. He further explains, "Besides, sexual purity in or out of marriage isn't a one-time vow, Grace. It's a daily recom-mitment to God and his plan for us." Thus, Jack serves as a narrative mouth-piece for Christian abstinence. Grace claims the ring and pledge are a "promise to them" when reassuring Jack about their relationship. This argu-ment, however, is an interesting frame deviating from what would be con-sidered a typical motivation for such a pledge. Purity pledges are traditionally promise to God and not to parents or boyfriends.

Attempting to pressure Grace into any form of sexual activity, Jack asks if oral sex is allowed, or if that "might even be a sin after you're married."[39] Repeatedly, and across episodes, Jack tries to pressure Grace to have sex, but when she does not give in, he searches for other means to lose his virginity. Ultimately, Jack becomes frustrated and begins cheating on Grace with Adrian. When Grace finds out about his sexual activity with Adrian, she pun-ishes Jack by ending the relationship. However, it is not long after the break up that Grace begins to rethink her pledge.

Shifting the accountability for her purity pledge away from her higher power to her parents and her relationship with Jack explains how it ultimately became easy for Grace to break her pledge. Valuing her boyfriend over the higher power, coupled with discovering her parents did not wait until mar-riage to have sex, Grace begins questioning the real purpose of the purity ring. In episode ten, "Back to School Special," Grace takes off her promise ring.[40] Explaining why she stopped wearing the ring, Grace says, "You know what, all of this started when I put on this ring, so I'm going to take it off." Jack offers an excited "alright," and Grace says, "I'm taking it off because I don't need it. I'm not tempted to have sex with you anymore. I'm not tempted to have sex with anyone." Finishing the story arc, in season two, Grace does finally have sex with Jack. Continuing the punishment framework, Grace experiences the most tragic punishment of all. Moments after her first sexual experience, it is revealed that her father has died in a plane crash. Immediately, the viewer observes Grace blaming herself for her father's death because she had sex, furthering the tragic frame for sexual relations for unwed teens.

Although the majority of the lessons presented in the show enforce tangible redemption for positive behaviors and condemnation for poor choices, the fact that open conversations about sex are occurring should not go unnoticed. Unlike most shows, *The Secret Life of the American Teenager* provides countless discussions about sex, occurring between partners, peers, even parents. Though less frequent, numerous mentions of condoms are found throughout the show. Unfortunately, the emphasis far too commonly centers on the prevention of pregnancy and not on prevention of sexually transmitted illnesses or HIV. Many of the conversations are also full of misguided information. For instance, Ricky tells Adrian that the stopping and starting of sex hurts, and that it is "not healthy for a guy." Continuing his cautionary tale, he recounts a story of a guy who had to go to the hospital and suffered permanent damage and is now sterile.[41]

All in all, *The Secret Life of the American Teenager* offers viewers models for a number of vital conversations. The show presents condom negotiations between partners, interactions with parents and teens about sex and pregnancy, and even highlights friends discussing fears and pressure to give in to a partner. However, these discussions occur through a tone seeped with overt Christian overtones. In fact, most of the episodes feel like the afterschool specials of the early nineties, with an overly didactic, moralizing tone. Even so, the show remained a popular draw for younger viewers until its cancellation in 2013.

In stark contrast to these two shows is *Pretty Little Liars*. While the show shares a high school setting with *The Secret Life of the American Teenager*, much of the overlap ends there. Little to no discussion of faith or abstinence occurs within *Pretty Little Liars*. Similarly, discussions about condoms or any means of protection are absent from the narratives within the show. Instead, *Pretty Little Liars* tells the twisted and drama-filled story of four female friends: Aria (Lucy Hale), Spencer (Troian Bellisario), Hanna (Ashley Benson), and Emily (Shay Mitchell). The crux of the show is a mystery of what happened to the girls' friend, Alison (Sasha Pieterse), and someone who blackmails the group of girls under the pseudonym of A.

Differentiating itself further from *Secret Life*, *Pretty Little Liars* is notable for its inclusion of a lesbian couple. While not a primary plot line of the show, the process of Emily discovering her sexual orientation and coming out provides a significant lesson for viewers. From the beginning, A has been blackmailing Emily for secretly being a lesbian. Though it is implied that Emily had fallen in love with Alison prior to her death, Emily is initially shown having a boyfriend. Emily's storyline provides viewers the opportunity to consider that it is common to be confused about one's sexuality.

Emily is able to see the truth for herself once she meets her first girlfriend, Maya (Bianca Lawson). Keeping her relationship with Maya a secret for most of the first season, she finally breaks the news to her friends in episode nine, "The Perfect Storm." Discovering Emily's sexuality, Hanna quickly urges Emily to tell the whole group, reassuring Emily that she can trust them. Still experiencing the very real concerns about coming out to family and friends, Emily chooses to wait. Emily's coming out process then becomes a realistic model of the struggles many young LGBTQ teens experience. When the news finally breaks, Aria and the others openly come to Emily's side, explaining they wish they had known sooner.[42]

Emily's parents, on the other hand, possess strong conservative values and struggle a great deal more with Emily's coming out process. Trying to support their daughter, the family invites Maya over for dinner. Unfortunately, after the dinner Emily's mother directly tells Emily, "I'm not okay with it. The whole thing makes me sick to my stomach."[43] While this show demonstrates progress through the inclusion of the young homosexuals, Christian undertones that dominated earlier shows are still demonstrated in Emily's mother's initial reaction to her daughter's coming out. Her parents do eventually support Emily more in later seasons of the show.

Echoing *Greek*'s significant emphasis on sexual activity, *Pretty Little Liars* also portrays a hookup culture, however this time among the high schoolers. The young characters are shown frequently lying about who they are and about their age in order gain the attention of older characters. In the pilot, for instance, after implying that she is a student at the local college, Aria is shown making out with a man she just met in bar after discussing a shared passion for writing.[44] She later discovers that the man is her new English teacher, Ezra (Ian Harding).

Each of the four primary characters in the show engages in some form of sexual relationship. Although it is common for young people to be sexually active in high school, the reliance on deception to maintain each of the relationships is troubling. Equally disturbing is the amount of infidelity and shaming of women for being sexually active. A perfect example of "slut shaming" is found when Spencer is accused of having slept with most of her older sister's boyfriends. While these accusations provide a glimpse into the sisters' history and explain the rift observed between Spencer and her sister, they also serve as a way to shame Spencer for any perceived acts of promiscuity, which is never confirmed or denied with any certainty. While there are plot devices that problematize the situations concerning sex, the act itself is shown throughout the show as normal behavior. Beyond Emily's struggle to accept her identity, sex is simply something that just happens. In many ways, this normalization of sex can be seen as positive progress, and a closer depiction of the lived reality of high school students. That said, given the age of the characters, and viewers alike, *Pretty Little Liars*, even more than others on ABC Family/Freeform, misses significant opportunities to engage the audience in the realities and consequences of having sex, including new perceptions of self, STIs, and other changes in the relationship dynamic.

Offering the greatest level of progress in both the inclusion of healthy relationships and open discussions about sex is the show *The Fosters* (2013–2018). Inviting audiences into the home of lesbian couple Lena (Sherri Saum) and Stef (Teri Polo) and their adopted and biological children, the show provides far more realistic lessons about sex than any of the shows previously discussed. From the outset of the series, the mothers openly discuss sex with their children, and early in season one, Stef makes sure her son Brandon (David Lambert) has condoms, since she knows he is sexually active with his girlfriend. As the show progresses, several other sexual health issues rise to the surface, including the morning after pill, sexually transmitted infection testing, sexual assault, and the concept of "friends with benefits."

One episode directly tackling sexual health is episode five of season one, "The Morning After." In the episode, Jesus's (Noah Centineo) girlfriend, Lexi (Bianca A. Santos), wants the morning after pill having unprotected sex.[45] Catching Jesus paying a woman to buy the morning after pill, Stef confronts the teens. Jesus admits that what they did was stupid, but that Lexi thought it was "the only thing to do" because "she is afraid her parents are going to find out and freak." Seeking to help, Lena and Stef have Lexi's parents over to discuss the relationship between their children. Upon discovering that their daughter has had sex with Jesus, Lexi's parents forbid her from continuing to see him. Although the young couple does continue to date, they choose to forgo any additional sexual activity.

Besides the more typical conversations about when a character should engage in sexual behavior, the first season of *The Fosters* also included a story line regarding sexual assault. Deepening an already complicated issue, the show places the topic within the foster care system. Stef and Lena's new foster daughter, Callie Jacob (Maia Mitchell), confides to her foster brother Brandon, that she was sexually assaulted in a former foster house. Without hesitation, Brandon encourages Callie to tell their mothers. Once she confides in her foster mothers about the abuse, they assist her in seeking help and encourage her to press charges, making sure that no future young girls will suffer sexual assault by the same man while in foster care.

Additionally, the high school characters are shown discussing sexually transmitted infection testing in several episodes, with Lena and Stef providing information to their children. But Lena and Steph are the only parents shown having these discussions with their children; other parents avoid the topic of sex in the show, aside from discussions of abstinence-only. Thus, Lena and Steph are the exception in open discussions about safe sex and the potential risks involved in being sexually active at a young age. Such open and informative conversations are a rarity, compared to other shows examined on ABC Family or otherwise. Lastly, the concept of "friends with benefits" is discussed when Jesus breaks up with his long-distance girlfriend, Lexi, as he asks another girl to engage in sexual activity with him, but only as a friend. While they do not pursue this type of relationship in the first season, they openly discuss it as an option.

Throughout the first season of *The Fosters*, the characters openly discuss sex and several of the consequences of being sexually active. While the topic of sex is prevalent in the show, the act of engaging in the behavior is rarely seen. The show is also far more progressive than the previous ABC Family/ Freeform shows in regards to how it approaches discussions of sexual orientation. The fact that *The Fosters* portrays a lesbian couple raising children in a very stable household indicates that writers have strategically chosen to demonstrate that nontraditional families can be as effective as traditional families.

Recovery Road (2016) is the most recently-aired show on ABC Family/ Freeform of the four shows examined in this investigation. The show takes on issues of sex, drugs, and the recovery process. Viewers are introduced to seventeen-year-old Maddie (Jessica Sula) as she drinks, smokes, and parties her way into an outpatient drug and alcohol rehab clinic. Focusing on Maddie's experience through 90 days of a mandated rehab program, including school by day and in-house counseling each night, the primary focus of the show is a tragic frame, highlighting the very real consequences of drugs and alcohol. Presenting a cast of other recovering teen and adult addicts, the show's central plotline illustrates many of the actual situations and feelings related with recovery, including denial, anger, lashing out, and facing the reality that sometimes "fun" actions lead to serious consequences.

One of the many realities Maddie faces in episode one is when her mother hands her a condom wrapper that she found in the car.[46] "I uh, found this [holds up condom wrapper] on the front seat of your car. I am glad you and Zack are being safe." What her mother doesn't know is that Maddie wasn't safe with Zack (Keith Powers). She tells her roommate, Trish (Kyla Pratt), "I don't remember. I don't remember anything from Thursday night. I have never had sex before. And if I did, it wasn't with my boyfriend, because he was babysitting his brother on Thursday." Trying to offer help, Trish reminds Maddie that if she doesn't remember the night, it is possible she didn't give consent. This raises the possibility of rape, which could be construed as Maddie's punishment in the tragic frame for being intoxicated. While the potential consequences here might seem regressive and incredibly extreme, the fact that Maddie's mother talks openly with her about the condom and safe sex is very progressive.

As the season moves forward, romantic tensions between Maddie and Wes (Sebastian de Souza), another teen in the house, begin to increase. One of the main rules in recovery, and in Maddie's sober home, is that participants are not supposed to be in a relationship for the first year. If someone is already in a relationship, he/she is not supposed to make any big changes to the relationship for a year. For Maddie, both rules are broken as she is in a relationship with someone else, Zack, yet begins a flirtation with Wes. In episode three, Wes explains, "It's all jumbled in my brain. The sex, the drugs, the love."[47] For Wes, it is not uncommon to replace one addiction for another, and this reality provides yet one more window into the rehab narrative.

Compared to the shows discussed previously, sex is far less prevalent in *Recovery Road*. In place of overt sexual activity, the show leans on creating drama through creating sexual tension. For example, tensions run high for Wes and Maddie, and eventually they do hook up. However, since the show was not given a second season, the possibility of the consequences of sexual activity outside the rules of the sober house are left to the imagination of the viewers. Additionally, the show also stirs up drama between two of the house's adult counselors, Craig (David Witts) and Charlotte (Sharon Leal), who is engaged to another man. In this show, sex might take a backseat to the other topics at hand; however, the lessons presented about sex being addictive as a drug and the temptation to cheat on a spouse still speak volumes.

As with many of the other shows on ABC Family/Freeform, faith and belief in God play a central role. The school counselor plays Christian pop music while driving Maddie to and from school, prayer is mentioned multiple times, and placing faith in something bigger than one's self is noted as a key to recovery. Although faith is discussed throughout, the tone is far less intense than in the early episodes of shows such as *The Secret Life of the American Teenager*. The discussion of faith in this context also make the most sense in terms of advancing the story. Faith in a higher power is a primary tenet of recovery programs[48]; thus, counseling sessions that mention God are to be expected if the program is to be considered realistic. That said, faith in relation to sex is completely absent. Nowhere in the show is the discussion of sexual activity ever discussed from a frame of faith.

Overall, *Recovery Road* highlights many real and relatable situations teens are facing today. While sexual health is only an underlying theme of the show, the awkwardness of dating, challenges of remaining faithful, and the temptation of cheating are clear. Overall, the show provided many important lessons for young viewers about the very real consequences of risky behavior.

Implications

As ABC Family/Freeform has developed over the years its portrayal of sexual relationships has evolved by offering more progressive examples in terms of both storylines and characters. Over the years, the channel has shifted away from the Christian ideals of abstinence-preferred messages and heteronormative relationships. In early shows, like *Greek* and *Secret Life*, characters are punished for sexual behaviors outside the bounds of marriage. Despite this, in all of the shows examined, sex is shown significantly impacting interpersonal relationships, much as it does in life. Often the most consistent form of punishment throughout the evolution of ABC Family/ Freeform's teen shows is the dissolving of relationships after sexual acts; however, not all relationships were negatively impacted by sexual activity. Surprisingly, nearly all of the shows, with the exception of *Pretty Little Liars,* offered at least one reference to the importance of protected sex, with the most common discussion of protected sex focused on avoiding pregnancy, rather than prevention of sexually transmitted infections. Overall, given that the shows' characters are teenagers, the only show that has what could be considered explicit depictions of sex, at least in the first season, is *Pretty Little Liars*. Despite *Pretty Little Liars'* more graphic depiction of sex, there is never a discussion of condoms or the possible consequences for having unprotected sex; however, there was one discussion of abstinence.

Ultimately, ABC Family/Freeform shows have become more progressive in their presentations of sex. The reliance on tragic frames as a narrative structure in shows such as *Greek* and *The Secret Life of an American Teenager* ended up offering fewer opportunities for reform than the shows employing comic frames, such as *The Fosters*. Perhaps the most dire punishments in early shows, like *Greek* and *The Secret Life of an American Teenager*, were reserved for those who violate the Christian values of the show. By contrast, later shows normalize and humanize LGBTQ characters, especially in *The Fosters* and *Pretty Little Liars*. But within the LGBTQ storylines, the discussion of sexually transmitted infections and use of condoms is markedly absent in the shows' first seasons.

As ABC Family has evolved and the network has rebranded to Freeform, it has relied less on religious beliefs to drive the narrative of the shows. This shift away from direct Christian discussions or proselytizing has meant that more recent shows no longer include purity pledges or abstinence. Additionally, the newer shows have lost the didactic tone that had become deeply threaded in every episode of *The Secret Life of an American Teenager*. In place of the educational and religious tropes, the channel's inclusion of sexual topics are replaced with more lifelike examples, working to advance the plot and drama of the shows.

Recent research claims that there is little to no correlation between watching shows with sex and young viewers engaging in sexual activity.[49] While viewing these shows may not lead to an increase in sexual activity, the pertinent question is whether young viewers see healthy relationships and sexual behavior. The programs discussed in this essay demonstrate the sexual act and the tensions surrounding sex, but not the potential life-long consequences of sex, except in the case of pregnancy in *The Secret Life of the American Teenager*, and they largely ignore other sexual health issues. More studies are needed to analyze the types of narratives and the behaviors that shows are promoting among young adults. Regardless of how unrealistic a show is, or how similar a viewer is to the character, one fact remains: talking about sex and sexual health continue to be perceived as difficult topics of conversation off screen. Scholars also agree that sex is a personally and socially sensitive topic.[50] Although *The Secret Life of an American Teenager* offered viewers direct conversations about sexual health, framing the message within a Christian context relies on viewer sharing in that belief structure to be effective. Beyond concerns of the viewer's beliefs, the fact that discussions of sexual health are either seen as a tragic consequence or rarely mentioned at all feeds the fear and inability to have comfortable conversations with partners and peers about sex. Most commonly, it is only when a crisis has arrived that a conversation about sexual health included on a show. In such discussions, emotions are already high, and the damage would already be done; thus, mentions of prevention are irrelevant. If shows could build in sexual health messages as prevention and not reaction, viewers could then learn how to talk about consequences, but also the desire to protect one's own body, thus empowering the viewer advocate and protect themselves.

Charles Ingold argues that society has established norms about what is appropriate and moral regarding sex.[51] While television narratives have come a long way in terms of talking about sex, shifting from showing married couples sleeping in separate beds to strangers or coworkers finding the nearest coat room for a "quickie," many of the characters in these shows seem to have no problem talking about having sex, without any conversation of sexual health consequences. For some viewers, openness in discussing sex and sexual health topics will provide a valuable model for how they can talk about such things. However, not all viewers will feel comfortable discussing these topics with older adults. Nor can it be assumed that everyone will have people with the knowledge about sexual health issues to ensure someone to help them with their concerns. Thus, given that none of the characters in any of the shows are ever shown talking to a medical provider about sexual health topics, these shows still leave a great deal to be desired. Until sexually-related topics are not seen as taboo and difficult topics to talk about with peers, partners, as well as doctors, it seems likely that any amount of HIV or STI narratives within television shows will ever overcome the perception that sex is not to be discussed.

NOTES

1. Elizabeth M. Perse and Jennifer Lambe, *Media Effects and Society* (London: Routledge, 2016).

2. Walter R. Fisher, *Human Communication as Narration: Toward a Philosophy of Reason, Value, and Action* (Columbia: University of South Carolina Press, 1989).

3. Glenn G. Sparks, *Media Effects Research: A Basic Overview* (Boston: Nelson Education, 2015).

4. Diana Oblinger, "Boomers, Gen-Xers and Millennials: Understanding the New Students," *EDUCAUSE* (July/August 2003): 37–48.

5. Jane D. Brown, Kelly Ladin L'Engle, Carol J. Pardun, Guang Guo, Kristin Kenneavy, and Christine Jackson, "Sexy Media Matter: Exposure to Sexual Content in Music, Movies, Television, and Magazines Predicts Black and White Adolescents' Sexual Behavior," *Pediatrics* 117, no. 4 (2006): 1018–1027.

6. Dale Kunkel, Keren Eyal, Keli Finnerty, Erica Biely, and Edward Donnerstein, *Sex on TV 4: A Kaiser Family Foundation Report* (Menlo Park, CA: Kaiser Family Foundation, 2005).

7. *Ibid.*

8. Victor C. Strasburger, Marjorie J. Hogan, Deborah Ann Mulligan, Nusheen Ameenuddin, Dimitri A. Christakis, Corinn Cross, and Daniel B. Fagbuyi, "Children, Adolescents, and the Media," *Pediatrics* 132, no. 5 (2013): 958–961.

9. Joshua Meyrowitz, *No Sense of Place: The Impact of Electronic Media on Social Behavior* (New York: Oxford University Press, 1986), 238.

10. Tara Brabazon, "Book Review: *The Age of Television: Experiences and Theories*," *Leisure Studies* 29 (2010): 111–113.

11. Milly Buonanno, *The Age of Television: Experiences and Theories* (Chicago: Intellect, 2008).

12. Victoria O'Donnell, *Television Criticism* (Los Angeles: Sage, 2007).

13. William J. Brown, "Assessing Processes of Relational Involvement with Media Personas: Transportation, Parasocial Interaction, Identification and Worship," *National Communication Association Conference* (November 2011), 15–19; Charles Salmon, "Setting a Research Agenda: Entertainment-Education Conference." Centers for Disease Control and Prevention, Office of Communication, 2001.

14. *Ibid.*

15. Brown, "Assessing Processes of Relational Involvement."

16. Alan M. Rubin and Mary M. Step, "Impact of Motivation, Attraction, and Parasocial Interaction on Talk Radio Listening," *Journal of Broadcasting & Electronic Media* 44, no. 4 (2000): 635–654.

17. Charles Salmon, Kim Witte, and Lee Byoungkwan, "The Effectiveness of Entertainment-Education as Media Health Campaigns: The Effects of Entertainment Narrative and Identification on HIV/AIDS Preventive Behavior," Meeting of the International Communication Association (New York, 2005); Suruchi Sood, "Audience Involvement and Entertainment Education," *Communication Theory* 12, no. 2 (2002): 153–172.

18. John T. Cacioppo, Wendi L. Gardner, and Gary G. Berntson, "The Affect System Has Parallel and Integrative Processing Components: Form Follows Function," *Journal of Personality and Social Psychology* 76, no. 5 (1999): 839.

19. Hans Alves, Alex Koch, and Christian Unkelbach, "Why Good Is More Alike Than Bad: Processing Implications," *Trends in Cognitive Sciences* 21, no. 2 (2017): 69–79.

20. Sparks, *Media Effects Research*.

21. Sheila T. Murphy, Lauren B. Frank, Meghan B. Moran, and Paula Patnoe-Woodley, "Involved, Transported, or Emotional? Exploring the Determinants of Change in Knowledge, Attitudes, and Behavior in Entertainment-Education," *Journal of Communication* 61, no. 3 (2011): 407–431.

22. Kenneth Burke, *Attitudes Toward History* (Berkeley: University of California Press, 1984).

23. Barry Brummett, "Burkean Comedy and Tragedy, Illustrated in Reactions to the Arrest of John DeLorean," *Communication Studies* 35, no. 4 (1984): 220.

24. "The Rusty Nail," *Greek*, season 1, episode 3, ABC Family, July 23, 2007.

25. "Liquid Courage," *Greek*, season 1, episode 5, ABC Family, August 6, 2007.

26. "The Rusty Nail," *Greek*.

27. *Ibid.*

28. *Ibid.*

29. "Depth Perception," *Greek*, season 1, episode 9, ABC Family, September 3, 2007.

30. "Picking Teams," *Greek*, season 1, episode 4, ABC Family, July 30, 2007.

31. *Ibid.*

32. *Ibid.*

33. "Black, White, and Read All Over," *Greek*, season 1, episode 10, ABC Family, September 10, 2007.

34. Laura Hamilton, "Trading on Heterosexuality: College Women's Gender Strategies and Homophobia," *Gender & Society* 21, no. 2 (2007): 145–172; Grahaeme A. Hesp and Jeffrey S. Brooks, "Heterosexism and Homophobia on Fraternity Row: A Case Study of a College Fraternity Community," *Journal of LGBT Youth* 6, no. 4 (2009): 395–415.

35. "Black, White, and Read All Over," Greek.

36. "Falling in Love," The Secret Life of the American Teenager, season 1, episode 1, ABC Family, July 1, 2008.

37. *Ibid.*

38. *Ibid.*

39. *Ibid.*

40. "Back to School Special," *The Secret Life of the American Teenager*, season 1, episode 10, ABC Family, September 2, 2008.

41. "Falling in Love," *The Secret Life of the American Teenager*.

42. "The Perfect Storm," *Pretty Little Liars*, season 1, episode 9, ABC Family, August 3, 2010.

43. "Salt Meets Wound," *Pretty Little Liars*, season 1, episode 12, ABC Family, January 10, 2011.

44. "Pilot," *Pretty Little Liars*, season 1, episode 10, ABC Family, June 8, 2010.

45. "The Morning After," *The Fosters*, season 1, episode 5, ABC Family, July 1, 2013.

46. "Blackout," *Recovery Road*, season 1, episode 1, ABC Family, January 2016.

47. "Surrender," *Recovery Road*, season 1, episode 3, ABC Family, February 8, 2016.

48. Kathleen M. Tangenberg, "Twelve-Step Programs and Faith-Based Recovery: Research Controversies, Provider Perspectives, and Practice Implications," *Journal of Evidence-Based Social Work* 2, no. 1–2 (2005): 19–40.

49. Christopher J. Ferguson, Rune K.L. Nielsen, and Patrick M. Markey, "Does Sexy Media Promote Teen Sex? A Meta-Analytic and Methodological Review," *Psychiatric Quarterly* 88, no. 2 (2016): 1–10.

50. Charles H. Ingold, "Socially Desirable Responding and Self-Reported Reactions to Sex on Television," *Southwestern Mass Communication Journal* 26, no. 1 (2010).

51. *Ibid.*

"You did this to yourself!"

Evaluating the Construction of the Fat Teenager in Huge

Andi McClanahan

The images are clear—bodies of fat teenagers on display. There are close-up shots of fat teens lying around in the grass, of dimpled legs and bulging bellies and robust arms, and of a fat young man lying on a bench with only his bare stomach viewable. One would believe they were watching one of the news reports about the "obesity epidemic" that have become so frequent in our society. However, in June 2010, this was the opening of the ABC Family's (now Freeform) fictional television series *Huge*.

Huge, a television show based on Sasha Paley's book by the same name, premiered in summer 2010. The show focuses on a group of overweight and obese teenagers attending Camp Victory, or as the general population calls it, "Fat Camp." *Huge* is unique in that it was the first fictional television show to be centered on overweight and obese teenagers. While there are reality shows and films such as *Too Fat for 15* (Style, 2010–2011), *Fat Camp* (MTV, 2006), and *Return to Fat Camp* (MTV, 2010) that shed light on this population, *Huge* is the first scripted show to deal with the issue of childhood obesity.

According to the Centers for Disease Control and Prevention,[1] nearly one in three children and adolescents are considered overweight or obese.[2] As a show that aimed to relate to a significant portion of the population, *Huge* should have been a hit. Harvey Guillen, who starred in *Huge* as Allistair, explains the excitement of the series: "It's going to be amazing—it has never been done before. We've already had so many people say that they can't wait for the show because there's a large percentage of teens in America who are overweight and they can't associate themselves with anyone on TV. They can't look at anyone and be like 'Wow, they look like me.' Everyone they see on TV is thin and glamorous and no one is being represented from the plus-

214

sized world. So there's a lot of teens out there who are totally going to connect with this and it doesn't have to be someone who is overweight because the storylines are so well written. I think it's going to be fantastic and people will want to watch these characters grow."[3] The series was highly anticipated because it was going to present something antithetical to what audiences are used to seeing on television—the thin teenager.

Huge started off strong for ABC Family, garnering "their biggest series launch to date [2010] among women 18–49 and second largest in adults 18–49 and women 18–34."[4] Yet, the ratings were not sustainable through the ten-episode run. The series lost viewers each week, with the finale having more than one million fewer viewers than for the series premiere.[5] The decision to cancel *Huge* was attributed to the low ratings for the series, which begs the question, why did people stop watching the show? When the news about the cancellation broke, Nikki Blonskey, who played the role of Willamena (Will), remarked, "We had a show that was so different. We were the first plus-size cast, I think, ever in Hollywood history. I just think it's kind of sad that TV stations are a little scared of having such a different show with such different people."[6]

In the summer of 2010, ABC Family released two other original programs—*Melissa and Joey* (2010–2015) and *Pretty Little Liars* (2010–2017). The other two shows had a cast that fit what is expected on television—all were thin and beautiful. *Huge* was the only one of the three shows to be canceled after the first season. The idea of representation in the media of individuals with different body shapes is important, but we also need to look at the messages that are created around those bodies. Media images and media narratives are powerful in constructing opinions and beliefs about others and ourselves. Douglas Kellner explains that various media forms help to "forge our very identities, including our sense of selfhood.... Media images help shape our view of the world and our deepest values.... Media spectacles demonstrate who has power and who is powerless."[7] In the case of *Huge*, viewers are shown what to think about individuals who are overweight or obese—including those individuals tuning into the show who are a part of this population. The media narrative of *Huge* must be interrogated due to the potential impact it has on younger people and how they approach these situations.

I argue that *Huge* failed in accurately representing and, ultimately, appealing to an audience because the show perpetuated the creation of the obese and overweight teenager as a spectacle—something to look at, something to pity, something to "other" and not want to be—and did little to change the hegemonic construction of fat or move away from the hegemonic ideal of thinness. Kellner explains that "media spectacle involves those media and artifacts that embody contemporary society's basic values and serve to enculturate individuals into its way of life."[8] In the case of *Huge*, the narrative serves to perpetuate the creation of spectacle as it relates to overweight and obese teenagers. Guy Debord argues, "when the real-world changes into simple images, simple images become real beings and effective motivations of a hypnotic behavior."[9] In other words, the continuous portrayal of overweight and obese individuals as problematic leads to an acceptance of these images and narratives without interrogation. Researchers have shown that images of the fat body have long been created as a media spectacle. However, while *Huge* attempted to move from the simple image of the fat body to constructing the narrative of the fat body and what it means to be an overweight or obese teenager, it did not change the master narrative but colluded with it.

In approaching *Huge*, I am undertaking a critical media analysis. I am studying the text from a humanistic perspective that allows me to emphasize "self-reflection, critical citizenship, democratic principles, and humane education."[10] Further, I am engaging in the text of *Huge* with the ambition of social justice. This involves a "desire to better our social world."[11] It is only through examining a text and analyzing how it works to perpetuate master narratives and ignore other possibilities that we can begin to move from latent viewers of media to active viewers of media. This is especially important as more analysis is accomplished in fat studies and body acceptance movements. Without critically analyzing the constructed messages for the public, we are unable to see where there are missed opportunities for alternative narratives and adherence to the hegemonic ideal of thinness.

In this remainder of this essay, I interrogate the episodes of *Huge* to answer the following research question: How does *Huge* work to perpetuate hegemonic beliefs regarding overweight and obese individuals through the framing of the narrative? For the purposes of this analysis, Farrell Corcoran's explanation of hegemony is useful: "Hegemony implies the active engagement of individuals with the ideology of the dominant sectors of society and therefore active cooperation in their own domination."[12] Therefore, hegemony assumes an acceptance of the dominant narrative as correct. Further, James Lull argues that "information and entertainment technology is so thoroughly integrated into the everyday realities of modern societies, mass media's social influence is not always recognized, discussed, or criticized…. Hegemony, therefore, can easily go undetected."[13] In order to move to a more productive media narrative about overweight and obese individuals, we need to point out the potentially damaging images currently being portrayed.

The negative representations of obese and overweight individuals contribute to the idea that "fat is bad" and thus, the fat person is bad. Ester Rothblum and Sondra Solovay explain, "*Fat is bad*. Isn't it odd that people deeply divided on almost every important topic can so easily seemingly organically agree on the above assertion? Isn't it similarly strange that countries significantly divergent in culture, attitudes, and approaches apparently share the fat-is-bad sentiment?"[14] Some researchers argue that there is no such thing as a "fat person," only a "fat body," but in our society we have a tendency to equate corporeality with personality characteristics. Kathleen LeBesco and Jana Evans Braziel posit that "fat equals reckless excess, prodigality, indulgence, lack of restraint, violation of order and space, transgression of boundary."[15] When we see fat as "indulgence" and "lack of restraint," we are accepting the hegemonic beliefs regarding overweight and obese individuals. Further, we are assuming that the characteristics of the body can be read as personality flaws.

According to Sei-Hill Kim and L. Anne Willis, there are two key ways television news and newspapers frame the topic of obesity. The first is to look at the individual in terms of what the individual did to cause the problem and what can the individual do to fix it. The second approach takes a broader view in looking at how economic, political, and societal factors impact a problem. Their research discovered that most news stories on obesity only focused on personal responsibility.[16] LeBesco and Braziel also articulate that fat is considered an acceptable reason for discrimination based on the assumption of personal responsibility: "Terms deemed offensive to less powerful groups are frowned on. Racial and ethnic jokes are less frequently punctuated by laughter. Stories that denigrate women and physically challenged people are not well received. Still, there is something about fat that escapes this change. People openly, disparagingly refer to themselves and others as fat. Perhaps it is because fat is a subject-marking experience over which we are perceived to have some degree of control (unlike gender or race, which are commonly—though mistakenly—taken to be fixed, static identifiers) that fat continues to be so maligned."[17]

If one can control the deviant behavior—in this case being overweight or obese—then there is no reason to look at other causes of the deviance. Harvey Guillen explains in an interview for *Cambio* that the purpose of Camp Victory is to "help with healthy habits that help you get back on track with what weight you should be. It's working out five hours a day and watching what you eat."[18] It is all about what individuals can do to conform to what is expected in our society when it comes to body size and weight. The only ways to change your body size and shape presented in *Huge* are through food choices, food restriction, and exercise. In relation to hegemony, this leads to the belief that individuals must be disciplined by others and discipline themselves into the societal norms if they want to be treated equitably.

After analyzing the ten episodes of the first season of *Huge*, it is clear that the framing of the overweight and obese teenager embraces the hegemonic ideal of thinness. *Huge* accomplishes this in multiple ways. The first is through the construction of hierarchies of thin versus fat. Specifically, *Huge* exposes how fat teenagers rank in regards to popularity in society but also how hierarchy operates even among a group of overweight and obese teenagers by placing the thinnest of the fat teenagers as the most popular. With the ultimate goal of rising up the hierarchy, *Huge* shows how the campers at Camp Victory are subjected to external disciplinarians while attending camp in order to shed weight and alter their body size. As external disciplinarians are not always the most effective in changing behavior, *Huge* also demonstrates how the campers discipline themselves in order to try to fall into the hegemonic ideal of thinness and thus reach a higher status in their social world.

Before diving into the analysis of the characters in relation to external disciplinarians and self-discipline, we must recognize that *Huge* shows the hierarchy that exists when it comes to thin versus fat individuals. Michel Foucault explains, "Discipline is an art of rank, a technique for the transformation of arrangements. It individualizes bodies by a location that does not give them a fixed position, but distributes them and circulates them in a network of relations."[19] Therefore, to view the fat body, we do so in relation to other bodies—often those that are thinner. When it comes to corporeality, the fat body ranks lower than the thin one in acceptability. This often leads to overweight and obese individuals attempting to discipline their bodies to climb the hierarchy and become more acceptable. Further, individuals often will succumb to external disciplinarians in order to help them transform into something more acceptable—the thin body. Elaborating, Le'a Kent explains, "In the public sphere, fat bodies … are represented as a kind of abject: that which must be expelled to make all other bodily representations and functions, even life itself, possible."[20] Transforming the fat body becomes the most important task for those who are attempting to fit into societal norms. When speaking of the media influence on attitudes towards the fat body, Dina Giovanelli and Stephen Ostertag argue, "viewers are simultaneously reminded that violating expectations of physical appearance, perhaps by being fat and female, will be recognized and subject to gossip and discrimination."[21] After watching shows such as *Huge*, viewers are reminded that they must have a body that fits into the hegemonic ideal in order to avoid ridicule and isolation.

The fact that overweight and obese teens must sequester themselves at a camp for a summer away from the rest of society to accomplish their goals of losing weight and becoming "normal" presents the campers as deviant as it relates to the hegemonic ideal of thinness. They are automatically ranked lower on the hierarchy when it comes to corporeal acceptability. Statements are made that show being at Camp Victory is considered being a part of a lower class of standards when it comes to friendship and romance. Becca (Raven Goodwin) says to Will on the first day, "It's not that bad here…. You meet people. People hook up. I mean, not everyone. But, you see, everyone is overweight so the playing field is more like there is one."[22] Becca is making it clear to Will that while they would not stand a chance for love in the "real world," at Camp Victory things are different. Chloe (Ashley Holliday Tavarez) and Amber (Haley Hasselhoff) further demonstrate this notion of Camp Victory versus the "real world," when Chloe asks Amber if a male camper is cute. Amber replies with "cute for here," clearly showing that "here" isn't nearly as good as the outside world.[23]

While the viewer can deduce the construction of a hierarchy between Camp Victory and others, *Huge* takes it a step further and shows interactions that the campers at Camp Victory have with those at a neighboring tennis camp. When the tennis camp participants come over, an unnamed camper from Camp Victory says, "So, you finally decided to put down the cake. When you're in the shower and you look down, what can you even see? What are you going to do? Eat us?"[24] The encounters between the tennis campers and those at Camp Victory continue when they run into each other while the campers from Camp Victory are participating in live action role play. A boy from the tennis camp exclaims, "Dude, I bet the butter trolls are back. Why are you all dressed up like that? Is it fat Halloween?"[25]

Hierarchies in *Huge* do not just exist between Camp Victory and the outside world. There are hierarchies within Camp Victory with the thinnest and most athletic being more desirable to others at the camp. The first scene positions Amber at the top of the hierarchy as the thinnest camper at Camp Victory. When Becca sees Amber after telling Will that there is a chance for love at Camp Victory, she remarks, "Okay. I guess it's not that different from the real world."[26] When two other girls see Amber at the first weigh-in, they approach her:

KAITLIN: (to Amber) Hi! We hate you.
AMBER: Okay?
CHLOE: Cute suit. I tried that one but it gave me back rolls.
AMBER: Oh, I know.
CHLOE: Are you kidding? You're so thin!
KAITLIN: You're totally the thinnest girl here.
AMBER: That can't be true. I mean, look at me.
CHLOE: Look at me![27]

This display of hierarchies among the campers continue with Amber as she is mistaken for one of the campers at the tennis camp by the tennis campers, and she is with them as they disparage the Camp Victory campers. Amber does not step in—instead being happy she is mistaken for a "normal" teen.

While Amber is at the top of the hierarchy, Will is at the bottom. When Will begins to like Ian, she realizes very quickly that Ian (Ari Stidham) has a crush on Amber—like most of the boys at Camp Victory. Will is practically invisible to the men at the camp as anything other than as a friend. When Will is seen outside the boys' cabin, one exclaims, "Hey! There's a girl out there," and another responds, "Oh, it's just Will."[28] As one of the larger women at the camp and the most outspoken, Will is at the bottom of the hierarchy of popularity. She does not embrace the idea of a feminine body or of feminine actions. Will resists the three areas that construct the feminine body, as defined by Sandra Lee Bartky: "those that aim to produce a body of a certain size and general configuration; those that bring forth from this body a specific repertoire of gestures, postures, and movements; and those that are directed toward the display of this body as an ornamented surface."[29] Her inability to display a feminine body renders her as nothing more than a friend among the male campers.

The final demonstration of hierarchy within the camp is through the character of Chloe. Allistair is Chloe's twin brother, but she keeps this a secret from the fellow campers because she does not want to be judged as he is larger than she is and has a more feminine appearance than is socially acceptable for a young man. It is not until the end of the series when the parents come for the weekend that Chloe reveals this information to her fellow campers. Chloe allows her brother to be humiliated by other male campers when one is dared to kiss him because Chloe does not want to relinquish her status at the camp as one of the most popular campers.

In addition to her treatment of her brother, Chloe ignores Becca, who was her closest friend the previous year at Camp Victory. Through flashbacks, the audience sees how close Becca and Chloe were as campers. Yet, when they get to the camp for the second summer, Chloe ignores Becca in hopes of being a part of the more popular group of girls. On the first day, Becca looks at Chloe and once she catches her eye, Chloe immediately looks away. Becca looks disgusted as Chloe looks away and it is clear to the audience that Becca is feeling a level of abandonment from her friend. Another camper notices and asks Chloe, "What's her problem?" Chloe explains, "She is just jealous. Last year I was a size eighteen and when I started losing weight, people started hating me."[30] Becca confronts Chloe at the Spirit Quest regarding her behavior and Chloe admits, "I know, I'm a bitch. But I wanted to be someone different this year."[31] Chloe reveals the hierarchy and her need to be in a higher position than she was the previous year when she weighed more; she wanted to be more popular thus chose Amber as her closest friend and rejected Becca.

The hierarchies show the apparent desire of the campers to move out of their social positioning as "fat kids" into roles that are more socially acceptable. One of ways for the characters to achieve this goal is through others' disciplining of the overweight and obese characters. This can be likened to Foucault's conception of discipline and power in the form of the Panopticon, which is a physical prison structure that has a disciplinarian at the center in a tower—one that can see others at all times. The effect is "to induce in the inmate a sense of conscious and permanent visibility that assures the automatic functioning of power."[32] Camp Victory is a setting based on the need for external sources of discipline. The idea that overweight and obese teens must leave their regular lives and enter a program only populated with others who are overweight or obese to lose weight shows that people think these characters need disciplinarians in their lives to help them change their habits. In *Huge,* several people serve as disciplinarians of the campers—the counselors, Poppy (Zoe Jarman) and George (Zander Eckhouse), Shay (Tia Texada), the fitness instructor, and Dr. Dorothy Rand (Gina Torres), the director.

The opening scene of *Huge*'s first episode, "Hello, I Must Be Going," demonstrates the idea that the characters on the show are to blame for their weight problems. Specifically, the first weigh-in requires the campers to be in bathing suits. When Will is in line in her shorts and a t-shirt, she is told that "they don't let you do the picture with clothes on." Will continues to stand there with Becca until Dr. Rand, Camp Victory's director, comes over and confronts Will. Dr. Rand states, "This is the start of a very important journey and we need you to begin by taking an honest look at yourself."[33] The idea that one must be nearly naked to take a look at themselves feeds into the spectacle of the fat body and also of personal responsibility. The implication is that the campers did this to their bodies and they must be shown the repercussions—the fat body.

Later in the same episode, Poppy goes through all the female campers' belongings to make sure they do not have any food items from home that are forbidden at Camp Victory. In this scene, we see Poppy taking everything from candy to gum. At one point, Poppy even considers confiscating toothpicks, something Amber chews on constantly throughout the series. When the campers are introduced to Shay, one camper exclaims, "She lives to make people cry. I love her."[34] As the campers run on a trail through the woods, one sits down on a rock to take a break. Shay yells, "There is no 'I can't.' Now get up off of that rock and give me 20 jumping jacks."[35] At the end of the episode, Will sneaks off to a diner near the camp and orders fries. Unfortunately for Will, Dr. Rand is there. Dr. Rand's statement to Will is one that challenges her behavior and lets her know she has disappointed her already. She asks, "So you would rather risk your life than change your life? Not so much funny as sad."[36]

The discipline continues throughout the series with the exercise regime, the food distribution, and the statements made by campers to each other. One of the key modes of discipline and visibility is the weekly circle discussion. Will expresses her discomfort of the discussions when she says, "I just want to scream every time we are in that circle. No pressure as long as you're trying as hard as you can to shrink to an acceptable size."[37] Another element of external discipline is the weigh-in in the middle of the season. During the weigh-in, Dr. Rand attempts to be encouraging to the students with her statement, "These high expectations, whether they come from us or people in our lives can be destructive. All anyone should expect of you, all you should expect of yourselves is that you try your best."[38] While Dr. Rand tries to be encouraging, she also demands the campers be in swimsuits and on full bodily display during the weigh-in process—an experience that is humiliating for many at the camp. One camper asks, "Do we have to wear bathing suits again?" and another camper responds exasperatedly, "Every time."[39] Will remarks at the first weigh-in, "Can we take a moment to just ponder how sick this is? Get a bunch of fat kids to come in bathing suits and add cameras."[40]

The campers also enact discipline against each other when they do things that may hurt their progress to the ideal thin and healthy body. Amber reveals to Becca that she is the one who informed Dr. Rand that Caitlin (Molly Tarlov), a camper who was only shown in the first episode, was binging and purging. The result of Amber telling Dr. Rand is that Kaitlin is kicked out of the camp the first week. Becca then reveals to Amber that she is the one who told Dr. Rand that Will was selling candy to other campers at the beginning of the summer.[41] This leads to a meeting between Dr. Rand and Will early in the series where Dr. Rand questions whether Will really wants to be at the camp.[42]

The discipline does not just exist with the leaders and the campers but also between Shay and the chef, Joe Sosniak, known as Salty (Paul Dooley). Salty is Dr. Rand's father, and it is suspected that nepotism is how he was hired for the job. However, Shay criticizes the food, saying it tastes too good to not be made without using fat. Salty simply responds with, "I made these with olive oil instead of margarine. The good kind of fat." One of the reasons the viewer can assume Shay is critiquing the food is because Salty is a larger man with a protruding stomach—he is not the traditional picture of health. In this case, Shay believes it is her right to critique the chef because she is the fitness trainer. Yet, Salty also serves as a disciplinarian of the campers with his common statement of "No seconds!" as he is serving the food.

The campers discern early on who will act as less of a disciplinarian in their quest for thinness. At one point, a camper asks, "What is circuit training?" and the response is "Who cares? George lets you take a 15-minute bathroom break."[43] Poppy is also seen as a weak disciplinarian. While she takes the food away at the beginning of the series, past incidents where her campers got into trouble—either for food infractions or sexual relations—are brought up regularly. The viewers learn later in the series that Poppy attended the camp, which may be one of the reasons she is a bit less of a disciplinarian.

She tells George she was a camper and he remarks, "I can't believe you were ever overweight,"[44] because Poppy is a very thin woman.

Perhaps even more telling than the external disciplinarians about the hegemonic ideal of thinness is the fact that people are quick to beat themselves up for falling outside of what is expected for body size. In her article "Foucault, Feminization, and the Modernization of Patriarchal Power," Sandra Bartky deals explicitly with the idea of the disciplinarian moving from an external source to an internal one as we deal with societal pressures, saying that "power … seeks to transform the minds of those individuals who might be tempted to resist it, not merely to punish or imprison their own bodies. This requires … a finer control of the body's time and of its movements—a control that cannot be achieved without ceaseless surveillance."[45] She explains that individuals end up policing themselves as they realize they are constantly on display for others and this discipline does not stop because they fear punishment from others, as she elaborates, "Discipline can be institutionally unbound as well as institutionally bound."[46] Thus, we are subject to discipline within the confines of external disciplinarians but also internal discipline from ourselves. This is especially true when it comes to issues of corporeality.

Bartky deals with the topic of dieting and how it can be conceptualized as disciplining oneself to societal expectations. She explains, "Dieting disciplines the body's hungers: appetite must be monitored at all times and governed by an iron will. Since the innocent need of the organism for food will not be denied, the body becomes one's enemy, an alien being thwarting the disciplinary project."[47] While *Huge* includes female and male campers at Camp Victory, the focus on self-discipline is largely placed upon the women. There are a few instances of male campers deriding themselves for their size, but their participation in the camp is related more towards health and the ability to be active than dealing with beauty. The exception is Allistair, who expresses his femininity openly to other campers. During the episode "Spirit Quest," the campers are told to come up with a name they feel best represents themselves. Allistair is paired with Trent (Stefan Van Ray) in their quest— two individuals who seem to be the opposite of each other. Allistair chooses the name Athena as the name he feels represents who he really is in life.

> TRENT: Why did you choose the name Athena?
> ALLISTAIR: She's the goddess of wisdom. She's awesome.
> TRENT: Yeah but, I mean, it isn't exactly normal to pick a girl's name.
> ALLISTAIR: Why does it matter? I like the name. I don't know why I should have to choose a boy's name just because people expect me to. I'd rather do what I want.[48]

As the most openly feminine male camper at Camp Victory, his self-discipline is almost expected in terms of his wanting to be more feminine thus fitting into the expected norms for women in terms of body size and shape.

The female campers are constantly self-disciplining themselves when it comes to food choices and how their bodies look. Bartky argues that while women are disciplined by external forces to adhere to expectations of the feminine and feminine body, women also subject themselves to their own disciplinary techniques like strict diets and excessive exercise.[49] Participation in Camp Victory is a statement that the campers are there to discipline themselves throughout the summer to lose weight. They have separated themselves from anything (or anyone) who may impede their success. The only person they have to deal with at Camp Victory in terms of not succeeding is themselves.

We learn in the first episode that Amber is the hardest on herself when it comes to punishing herself for her size. Amber desperately wanted to attend Camp Victory—saving her money for an entire year to spend her summer there to lose weight. On the first day at the camp, Amber decorates her sleeping area with pictures of models from fashion magazines and claims they are her "thinspiration."[50] She believes that seeing images of the ideal body image will inspire her to stick with her diet and lose weight. When her friends point out that she is a slow eater, she explains, "I chew every bite 30 times. I've been dieting since I was ten. It is probably the only thing I'm good at."[51]

Amber's self-discipline is one of the common threads throughout the entire series. She is continuously shown looking in the mirror while holding in her stomach or manipulating her body to appear thinner. During the second weigh-in, Amber becomes upset when she finds out she only lost one pound. Dr. Rand points out to her, "Don't be discouraged. Every 'body' is different. Kids that are much bigger than you tend to lose the first pounds more quickly. Just remember, it's not a race."[52] While Dr. Rand is attempting to be encouraging to Amber and point out logical reasons why she may be losing weight at a lower rate than others at the camp, this does not stop Amber from becoming extremely upset. She lies to Chloe and tells her that she lost six pounds. Later in the evening, Amber is shown looking at herself in the mirror, pinching her fat, and critically analyzing her body. She begins to cry.[53]

Out of everyone at the camp, Amber is shown to be the most self-conscious about her size. Amber does not believe she is worthy of attention because of her weight. When she becomes infatuated with George, a camp counselor, she is certain that he does not like her because of her weight and not because of her age or the status of their positions at the camp. After George and Amber kiss multiple times, he attempts to put an end to their sneaking off and meeting because he sees it as unprofessional on his part. Amber is convinced it is because of her weight and even argues with George to tell her the truth. George exclaims, "Amber, any guy would be lucky to be with you!"[54]

Amber continues to self-discipline herself in the ways she eats and her continual use of a toothpick. Early in the series, the viewer discovers that Amber constantly chews a flavored toothpick to help discourage her from eating and thus be able to lose weight. When Amber's mother comes to visit during family weekend, we see that her mother is a key component of her stress about food. Her mother is extremely thin and proclaims to Amber upon seeing her, "Don't get too thin!"[55] The viewer can see the pressures from Amber's mother in that she views Amber as competition. Amber's mother does not necessarily want her to be successful in losing weight because Amber could end up being competition in terms of perceptions of attractiveness from others. Amber's mother spends her time during family weekend flirting with other parents as well as George. It is during the visit that we learn Amber has a secret stash of cookies in the laundry room. She runs off to get the cookies, only to find out they have been taken by someone else. Amber goes to Will and asks her to help her get some food. They end up breaking into the cafeteria and eating brownies. However, Will catches Amber chewing her food and spitting it out into napkins. Amber explains, "This way I can taste it without getting the calories."[56] Amber's reaction to her mother's visit shows how her home life serves as a trigger for her in terms of stress and emotional eating.

Throughout the series, Will serves as the camper who is sarcastic and unhappy to be attending Camp Victory. It is known early on that her parents sent her to the camp because they were unhappy with her weight. In the first episode, Will does a strip tease when she removes her t-shirt and shorts after she is told that she must only be in her swimsuit for the first weigh-in. She exclaims, "You know, this could be my summer to gain weight. I feel that inside me there is an even fatter person trying to get out."[57] At the first circle where campers speak about their feelings, Will reveals, "I'm down with my fat. Me and my fat are like BFFs. Everyone wants us to hate our bodies. Well, I refuse to."[58]

Yet, even with Will's self-professed love of her body, she is seen disciplining herself in multiple ways when it comes to food and the choices she makes. In the first episode, as mentioned, Will sneaks off to a nearby diner and orders a chocolate shake and fries. When she discovers that Dr. Rand is in the booth behind her, Will gets up to leave before she is noticed and the waitress asks Will, "I've got your fries here. Do you want them or not?" Will does not respond but Dr. Rand answers, "She does."[59] Will, who seemed so adamant about being sent home from Camp Victory, is forced to decide. She must discipline herself not to eat the food she ordered and prove she wants to be at Camp Victory. Dr. Rand confronts Will and asks her if she fears failing at the camp. Will responds, "I'm not scared. I just think everything you stand for is crap."[60] In this interaction, the viewer sees that Will *is* fearful of failing at the camp. Later at the cabin, Amber whispers to Will, "It's not that it gets easier. It's that you start wanting it."[61]

Throughout the series, we see Will exhibit some behaviors that make it seem that she begins to want to lose weight and get healthier—even if she remains vocally opposed to the purpose of the camp. Will begins to express a desire to potentially hook up with a guy at camp, and this leads to her being more active and even learning how to play basketball.[62] During "Movie Night," Will begins to put on makeup to make herself look attractive to Ian.[63] She shows that she does care about being feminine and attractive to others. She wrestles with her food choices and tries to make good decisions. During "Spirit Quest," she and Amber get lost, and Will sees a waitress throw away an entire box of donuts. When Amber decides to head back into the woods, Will climbs up to the dumpster and pulls a donut out of the garbage. The donut has coffee grounds and lettuce on it, but she still struggles with putting the donut down and not eating it. She looks to the sky and says, "Help me, please." After seeing a milk carton with the words "organic" and "nonfat" on it, she throws the donut back into the garbage bin, demonstrating her discipline when it comes to eating healthier.[64]

Surprisingly, one of the characters most focused on self-discipline is Dr. Rand. The viewer discovers that she is a compulsive overeater. She attends weekly meetings to help her maintain control of her food intake. She is very strict with when and where she eats. She tells her father, "I don't eat this late" when they are together at the diner.[65] On her first date with Wayne (Stoney Westmoreland), a local contractor doing work for the camp, they are seen eating fast food in his truck. She lets Wayne know that she "never does this."[66] At the end of the series, Wayne asks Dr. Rand about her self-discipline and how she feels about herself:

> WAYNE: What were you like when you were fat?
> DR. RAND: I hated myself.
> WAYNE: Less?
> DR. RAND: Less.
> WAYNE: And that's the big improvement? You hate yourself less?
> DR. RAND: Yeah.[67]

This conversation shows that even though Dr. Rand has been successful at disciplining herself into a more feminine body and the hegemonic ideal of thinness, she still hates herself, which leads her to continual discipline.

After critically analyzing *Huge*, there are several areas where one can see that ABC Family missed opportunities with the audience. *Huge* was anticipated by television viewers, as is evident in the ratings of the premiere episode. However, audiences quickly stopped watching the show when they realized that the plus-sized characters were constructed in a similar manner as they would be on a television show not centered on overweight or obese people. The characters still focus on food choices and exercise and on hating themselves for their disobedience to the hegemonic ideal of thinness. What could have been an empowering show for the fictional characters and the body positivity movement quickly fell into the master narrative showing hierarchy and personal responsibility demonstrated through external disciplinarians and self-discipline.

Huge embraces the common belief that weight loss and health is solely due to a person's individual choices and their desire to be accepted. However, other factors influence why a person may become overweight or obese, and these are largely ignored in *Huge*. External factors such as genetics, socioeconomic factors, or environment are hinted at as possibly having an impact on the campers but are not fully explored. For instance, in the episode "Letters Home," a young woman arrives at the camp to join for the summer. However, her family members—who are all overweight or obese—stay to make sure she acclimates properly to Camp Victory. The family is overbearing with their daughter, staying at the camp even when Dr. Rand attempts to talk to them about the importance of leaving their daughter in her care. When the mother is speaking to Will, she says that they did not see the need for their daughter to attend the camp because they think she is beautiful.[68] It is clear there are environmental and hereditary factors playing a role with this young woman, but the script does not explore these possibilities. Instead, the daughter has a panic attack and leaves the camp, ending the possibilities for delving deeper into the reasons for obesity. The issues of heredity and potential underlying medical issues that may cause one to be overweight or obese are ignored in order to adhere to the dominant narrative that places blame on those that are outside of what is expected for bodily appearance and size.

There are numerous opportunities where the writers could have explored alternative reasons behind obesity that do not deal with just food choices and exercise. Will's parents are strict and extremely healthy. She mentions to Amber that they do not even allow junk food in the house. It is clear to a viewer knowledgeable in research in health that Will may be rebelling against her family through her body. However, it is not discussed. Ian speaks openly about how his mother and father fight all the time. Ian tells Dr. Rand that he is nervous about his parents visiting during parent's weekend because at home "there is Facebook or cake."[69] This statement shows that Ian is dealing with emotional eating—something that is constructed because of his home environment. Yet, *Huge* fails to fully address the issue. One of the causes of Amber's obsession with her weight becomes apparent when her mother comes to visit. Her mother is extremely thin and continuously says things like "Don't get too thin" to Amber. These statements demonstrate that Amber's mother is in competition with her daughter for who is the thinnest, most beautiful, and therefore, most desirable to men.

Perhaps the best example of a missed opportunity in dealing with the issue of obesity is with Becca. Becca explains to Will on the first day, "Can I say something? It's not that bad here. This is my second time. I lost twenty-six and a half pounds. Then, I gained some of it back. Basically, all of it back."[70] This statement shows that there is more to losing weight than simply following a program. The audience may wonder what it is that caused Becca to gain the weight back. Out of the entire cast of female campers, she is one who is the least obsessive about her weight even though she works to become healthier. Her goal is health. She says to Will at the end of a hike, "It's amazing. I can tell that I'm getting in shape. Even this walk. I'm not getting so out of breath."[71] At the end of the series, the audience discovers that Becca's grandmother—her closest family member—died during the past year. Again, *Huge* missed an opportunity to deal with the impact stress and emotions can have on eating and exercising. The writers do little to look at issues of obesity on a macro level.

While *Huge* leaves a lot to be desired from the perspective of a critical media analysis, it is an important show on ABC Family in that it tried to add a narrative for a population of teens who are largely ignored. Despite the bulk of the narratives being about becoming thin and focusing on the micro level solutions to obesity, the audience did get to see relationships form between the campers that were based on friendship and the desire to be better. As individuals consider possibilities for future television shows dealing with overweight or obese teenagers (or people of any age), *Huge* serves as a good example of why we need to broaden the discussion about obesity from simply personal responsibility to one that includes issues of heredity, socioeconomic issues, or environmental factors. *Huge* did have moments that worked to challenge the hegemonic ideal of thinness and the need for external disciplinarians or self-discipline. Dr. Rand asks the campers at campfire, "Is it even possible to stop attacking other people and ourselves and just surrender?"[72] We can only hope.

Notes

1. While the statistics presented by the Centers for Disease Control and Prevention are often cited as evidence of the prevalence of overweight and obese teenagers, there are many who debate the definition of obesity as well as the repercussions of obesity. For a more nuanced explanation of this debate, see Linda Bacon, *Health at Every Size: The Surprising Truth About Your Weight* (Dallas: BenBella Books, 2008).

2. Cynthia L. Ogden, Margaret D. Carroll, Cheryl Dr. Fryer, and Katherine M. Fiegal, "Prevalence of Obesity Among Adults and Youth: United States, 2011–2014," *NCHS Data Brief, Centers for Disease Control and Prevention and National Center for Health Statistics* 219 (November 2015): 1–8.

3. Ashley Rose, "Read Our Q & A with Harvey Guillen from 'Huge,'" *Cambio*, June 24, 2010, http://www.cambio.com/2010/06/24/our-qanda-with-harvey-guillen-from-huge/.

4. Trevor Kimball, "*Huge*: ABC Family TV Show Cancelled; No Season Two," *TV Series Finale*, October 4, 2010, http://tvseriesfinale.com/tv-show/huge-canceled season-two-18740/.

5. *Ibid.*

6. Jennifer Armstrong, "'Huge' Cancellation: Nikki Blonsky Talks About the Surprise

Demise of Her ABC Family Show and the Importance of Plus-Size TV," *Entertainment Weekly*, October 13, 2010, http://ew.com/article/2010/10/13/huge-cancellation-nikki-blonsky-talks/.
    7. Douglas Kellner, "Media Culture and the Triumph of the Spectacle," *Douglas Kellner: Essays*, 1995, https://gseis.ucla.edu/faculty/kellner/essays/mediaculturetriumphspectacle.pdf.
    8. *Ibid.*, 3.
    9. Guy Debord, *Society of the Spectacle* (New York: Black and Pearl, 2000).
    10. Brian L. Ott and Robert Mack, *Critical Media Studies: An Introduction*, 2d ed. (New York: Wiley Blackwell, 2014), 16.
    11. *Ibid.*, 17.
    12. Farrell Corcoran, "Television as Ideological Apparatus: The Power and the Pleasure," *Critical Studies in Mass Communication* 1, no. 2 (1984): 142.
    13. James Lull, "Hegemony," in *Gender, Race, and Class in Media: A Critical Reader*, ed. Gail Dines (Los Angeles: Sage, 2015), 40.
    14. Ester Rothblum and Sondra Solovay, "Introduction," in *The Fat Studies Reader*, ed. Ester Rothblum and Sondra Solovay (New York: New York University Press, 2009), 2.
    15. Kathleen LeBesco and Jana Evans Braziel, "Editors' Introduction," in *Bodies Out of Bounds: Fatness and Transgression*, ed. Kathleen Lebesco and Jana Evans Braziel (Berkeley: University of California Press, 2001), 3.
    16. Sei-Hill Kim and L. Anne Willis, "Talking about Obesity: News Framing of Who is Responsible for Causing and Fixing the Problem," *Journal of Health Communication* 12, no. 4 (2007): 359–376.
    17. LeBesco and Braziel, "Editors' Introduction," 2.
    18. Rose, "Read Our Q & A."
    19. Michel Foucault, *Discipline and Punish: The Birth of the Prison* (New York: Vintage Books, 1979), 146.
    20. Le'a Kent, "Fighting Abjection," in *Bodies of Our Bounds: Fatness and Transgression*, ed. Jana Evans Braziel and Kathleen LeBesco (Berkeley: University of California Press, 2001), 135.
    21. Dina Giovanelli and Stephen Ostertag, "Controlling the Body: Media Representations, Body Size, and Self-Discipline," in *The Fat Studies Reader*, ed. Ester Rothblum and Sondra Solovay (New York: New York University Press, 2009), 290.
    22. "Hello, I Must Be Going," *Huge*, season 1, episode 1, ABC Family, June 28, 2010.
    23. "Movie Night," *Huge*, season 1, episode 5, ABC Family, July 26, 2010.
    24. "Live Action Role Play," *Huge*, season 1, episode 3, ABC Family, July 12, 2010.
    25. *Ibid.*
    26. "Hello, I Must Be Going," *Huge*.
    27. *Ibid.*
    28. "Movie Night," *Huge*.
    29. Sandra Lee Bartky, "Foucault, Femininity, and the Modernization of Patriarchal Power," in *The Politics of Women's Bodies: Sexuality, Appearance, and Behavior*, ed. Rose Weitz (New York: Oxford University Press, 1998), 27.
    30. "Hello, I Must Be Going," *Huge*.
    31. "Spirit Quest," *Huge*.
    32. Foucault, *Discipline and Punish*, 201.
    33. "Hello, I Must Be Going," *Huge*.
    34. *Ibid.*
    35. *Ibid.*
    36. *Ibid.*
    37. "Poker Face," *Huge*, season 1, episode 7, ABC Family, August 9, 2010.
    38. *Ibid.*
    39. *Ibid.*
    40. "Hello, I Must Be Going," *Huge*.
    41. "Spirit Quest," *Huge*.
    42. "Talent Night," *Huge*, season 1, episode 4, ABC Family, July 19, 2010.
    43. "Letters Home," *Huge*, season 1, episode 2, ABC Family, July 5, 2010.
    44. "Spirit Quest," *Huge*.

45. Bartky, "Foucault, Femininity, and the Modernization of Patriarchal Power," 40.
46. *Ibid.*, 36.
47. *Ibid.*, 27.
48. "Spirit Quest," *Huge.*
49. Bartky, "Foucault, Femininity, and the Modernization of Patriarchal Power," 36.
50. "Hello, I Must Be Going," *Huge.*
51. *Ibid.*
52. "Poker Face," *Huge.*
53. *Ibid.*
54. *Ibid.*
55. "Parent's Weekend Part One," *Huge*, season 1, episode 9, ABC Family, August 23, 2010.
56. "Parent's Weekend Part Two," *Huge,* season 1, episode 10, ABC Family, August 30, 2010.
57. "Hello, I Must Be Going," *Huge.*
58. *Ibid.*
59. *Ibid.*
60. *Ibid.*
61. *Ibid.*
62. "Letters Home," *Huge.*
63. "Movie Night," *Huge.*
64. "Spirit Quest," *Huge.*
65. "Hello, I Must Be Going," *Huge.*
66. "Birthdays," *Huge*, season 1, episode 8, ABC Family, August 16, 2010.
67. "Parent's Weekend Part Two," *Huge.*
68. "Letters Home," *Huge.*
69. "Parent's Weekend Part One," *Huge.*
70. "Hello, I Must Be Going," *Huge.*
71. "Poker Face," *Huge.*
72. "Live Action Role Play," *Huge.*

"Deaf is not a bad word"
The Positive Construction
of Disability in Switched at Birth[1]

Anelise Farris

With the popularity of video streaming services like Netflix and Hulu, and the luxuries of having DVR service, it is clear that television watching is a valued recreational activity; nevertheless, the consumption of media is not merely a passive experience. Media has the capacity to both actively shape and affirm worldviews as well as to challenge preconceived notions. This latter potential is of particular importance for marginalized groups, like the disabled community, who are often portrayed in the media as one-dimensional, token characters. Although shows like HBO's *Game of Thrones* (2011–present) and the many DC and Marvel television shows have presented audiences with popular disabled protagonists, these shows are not set in the "real" everyday world that we inhabit. And the latter in particular perpetuates the narrative that disabled people are *super*human—which only serves, in turn, to *de*-humanize them. Furthermore, while there are popular mainstream shows that positively depict disabled characters—like NBC's *Parenthood* (2010–2015) and ABC's *Speechless* (2016–present)—a persistent problem remains in the lack of fully-developed disabled characters in shows that are geared toward younger audiences. Arguably, simply for this reason alone, the young adult television show *Switched at Birth* (2011–2017), the highest-rated ABC Family series debut and winner of numerous awards (including a TCA Award in 2012 for "Outstanding Achievement in Youth Programming" and a Peabody Award in 2013), deserves attention.[2]

In this essay I examine the implications of the disability-positive show *Switched at Birth*, arguing that the airing of the show's final season on the newly named Freeform channel culminates the show's commitment to disrupting "safe" portrayals of disability—a bold, yet necessary move for positive representation of and response to difference, particularly among younger audiences.

ABC Family's *Switched at Birth* is an American television series that premiered on June 6, 2011, and its fifth and final season was broadcast on Freeform in 2017. Created by Lizzy Weiss, the series centers on the lives of two Kansas City, Missouri, teenagers, Bay Kennish (Vanessa Marano) and Daphne Vasquez (Katie Lynn Leclerc), who become intimately involved in each other's lives after learning that they were accidentally switched at birth. John and Kathryn Kennish (D.W. Moffett and Lea Thompson), their son Toby (Lucas Grabeel), and non-biological daughter Bay are a quintessential white, conservative, upper-class family—consistently using money and affluence in foolish and excessive ways. Daphne, the Kennish's biological daughter, is taken home by Regina Vasquez (Constance Marie) and Angelo Sorrento (Gilles Marini), though she is raised primarily by Regina and her mother Adrianna (Ivonne Coll). In stark contrast to the Kennish lifestyle, the Vasquezes are a Puerto Rican family, living paycheck-to-paycheck in East Riverside, who reluctantly move into the Kennishes' guesthouse so that the families can become acquainted with their biological daughters.

While the Kennishes and the Vasquezes are remarkably different, what resonates throughout the show is that difference in terms of ability, race, and class do not (if we do not let them) divide us as human beings. Ultimately, *Switched at Birth* is a coming-of-age story that depicts the difficult process of individuation and the gradual, yet rewarding, acceptance of self and Other. In a powerful, formative medium—like television—that is consumed by the average young adult on a daily basis, the wise pairing of the coming-of-age story with the diversity-positive message warrants attention. By effectively dismantling binaries of normal and abnormal, while not diminishing the disabled experience, positive media representation for young adults—such as occurs in *Switched at Birth*—serves to facilitate a more accepting and confidently diverse society.

This essay focuses on both the portrayal of disability in *Switched at Birth*, as well as the reception of and/or response to disability by audiences of the show. First, I contextualize my analysis by reviewing how disability has been previously represented on the ABC Family channel, and, building upon this groundwork, the crux of the essay highlights significant moments from the series that force viewers to reconsider their understanding of disability. Reflecting on what contributes to the success of *Switched at Birth* as a disability-positive show geared toward young adults, I draw particular attention to how disability intersects with matters of race, gender, and sexual orientation, as well as simply the everyday, lived realities of the characters. Finally, the latter portion of this essay examines the reception of *Switched at Birth* by both the disabled and non-disabled communities. I conclude by reiterating how *Switched at Birth* overcomes, in significant ways, a long history of unfair representation—with special emphasis on the importance of Freeform's target audience. Ultimately, I intend to illustrate the importance of what this channel is providing through shows like *Switched at Birth*: a series which not only offers complex disabled characters but also critically engages with matters of disability. Fortunately, the move from ABC Family to the name *Freeform* confirms that the channel will continue to offer viewers exactly what they need: a reality in which diversity is indeed the norm.

The Trouble with Family

For nearly thirty years, the channel has had *family* in the title. Founded by the Christian Broadcasting Network in 1977, the channel quickly became associated with conservative values and "wholesome" family-friendly content. The channel itself was declared The Family Channel in 1990 to better address the channel's target audience: families. Although the channel has continued to aim their broadcasts at families, it has undergone multiple changes in ownership: Fox in 1997 (Fox Family) and Disney in 2001 (ABC Family). Most recently, the channel has been renamed Freeform—which officially became the title of the still Disney-owned channel in January of 2016. For many viewers the name change is of little importance, as the channel will continue to broadcast many of the same shows. However, the name change does signal that fewer restrictions ("freer form") will be placed on the shows that the channel airs—an outcome which seems directly tied to their decision to remove the family marker.

When the word *family* is used for marketing purposes, as in the channel ABC Family, there is generally the assumption that the product is family-friendly or safe for all ages. While seemingly trivial in theory, this notion is complicated by the fact that the term *safe* is too often aligned with *normal*: the model nuclear family, gender norms, and perfect emotional, mental, and physical health. However, all of these "ideals" are no more than constructs that are illusory at best. Furthermore, positive representation of diversity is particularly crucial with respect to the audience that ABC Family markets: teenagers and older young adults. In fact, the name change itself was actually motivated in large part by audience. On the Freeform website, the FAQs read, "We wanted a name that felt more like us. We're changing our name to be a better fit for our shows and social platforms, which are more targeted to young adults than families."[3] It is not just young adults (13–17) that these shows are marketed to, but those older young adults and 20-somethings as well: "Becomers," as ABC Family President Tom Ascheim calls them.[4] Vanessa Marano, who plays Bay Kennish on *Switched at Birth*, confessed in an interview,

"Freeform means no boundaries—but the good kind of no boundaries, not the kind that leave you screaming, 'boundaries, dude,' when your roommate walks around your apartment without pants."[5] By removing *family* from the title, and choosing a name like Freeform, which suggests a willingness to traverse boundaries, the channel is helping to deconstruct binaries like safe and unsafe, normal and abnormal.

Before addressing *Switched at Birth* in particular, it is worth briefly considering a few of the ways ABC Family has presented viewers with a range of disabled characters over the past decade—some portrayed more positively than others. *The Secret Life of the American Teenager* (2008–2013), created by Brenda Hampton, is a television series that ran for five seasons. Tom Bowman, the adoptive son of Marshall and Kathleen Bowman (John Schneider and Josie Bissett) and adoptive brother to Grace (Megan Park), is played by Luke Zimmerman, an actor with Down Syndrome. Although Tom figures as comic relief in the earlier seasons, his character becomes increasingly complex. After his father passes away in a plane crash, Tom steps into his role as "man of the house," and yet, however loyal he is to his own family, he craves his independence and eventually moves into the family guesthouse. Tom is also sexually active and capable of holding a full-time job, eventually even becoming a vice president of human resources. Therefore, although his role is limited in the early seasons of the show, his character develops significantly—a move which suggested that the channel was moving in a more progressive direction in their depictions of disability.

Unfortunately, however, not all portrayals of disability on the channel have been so positively constructed. For example, the show *Pretty Little Liars* (2010–2017), created by I. Marlene King, and loosely based on the *Pretty Little Liars* novels by Sara Shepard, implements the disabled person-as-villain trope. In the show, Spencer (Troian Bellisario), Aria (Lucy Hale), Hanna (Ashley Benson), and Emily (Shay Mitchell) work to uncover the circumstances surrounding the mysterious death of their friend Alison (Sasha Pieterse). Initially one of the potential suspects is Jenna Marshall (Tammin Sursok): a blind character, in oversized movie-star sunglasses, who is angry, manipulative, and sexually deviant. Additionally, her blindness fluctuates throughout the seasons: she has surgery in season two; is able to see in season three; and is blind once again in season four. Regrettably, her blindness appears to be nothing more than a dramatic device to further the plot.

Likewise, although the show *The Fosters* (2013–2018), created by Peter Paige and Bradley Bredeweg, is full of diverse characters in terms of sexual orientation and race, disability is surprisingly absent from the show—with the exception of Jesus Adams Foster (Jake T. Austin and Noah Centineo), the adoptive son of Stef and Lena Adams Foster (Teri Polo and Sherri Saum). Jesus, who has ADHD, regularly takes Ritalin for his disorder. If he does not take his medication, he becomes aggressive and has trouble concentrating. However, generally speaking, Jesus's disability narrative is more implied than explicit, and it often comes across as an afterthought or forced character development. While certainly the presence of ADHD in *The Fosters* is better than a total absence of disability, I maintain that invisible disabilities deserve more constructive representation in television programs. As shows begin to air on Freeform—no longer restricted by the illusion of family-friendly normalcy—I am hopeful that writers will take note of how ability diversity is as important as racial and sexual diversity and, consequently, deserves to be actively included in their programming.

The Politics of Naming

For the purpose of this essay, I will be using the terms *disabled* and *nondisabled.* Although I do realize that this contributes to a binaried discourse, this is not my intention, and, as such, I will briefly explain my reasoning. Author and advocate Nancy Mairs boldly states, "I am a cripple. I choose this word to name me. I choose from among several possibilities, the most common of which are 'handicapped' and 'disabled.'"[6] Mairs opposes "differently abled," which, as she writes, "partakes of the same semantic hopefulness that transformed countries from 'undeveloped' to 'underdeveloped,' then to 'less developed,' and finally to 'developing' nations. People have continued to starve in those countries during the shift. Some realities do not obey the dictates of language."[7] As Mairs observes, while language is a powerful tool, it can be as limiting as it is affecting. Disability studies scholar Simi Linton acknowledges that there are "nice words" such as *physically challenged,* "nasty words" like *retard,* and those used by the disabled community because of their "transgressive potential": words like *cripple* and *freak.*[8] And, while it is obvious enough to choose the "nice words" over the "nasty words," what is less apparent is which "nice word" to choose.

Due to this dilemma, some have chosen to use the terms *disabled* and *temporarily able-bodied* (TAB) as opposed to *disabled* and *nondisabled.* Brenda Jo Brueggemann, in an effort to emphasize the pervasiveness of disability, asserts, "As the saying goes in disability circles these days: 'If we all live long enough, we'll all be disabled. We are all TABs—temporarily able-bodied.' We are as invisible as we are visible."[9] The notion of TAB contextualizes everyone within the realm of disability—removing the need for "us versus them" discourse. Yet, by choosing to use *TAB* over *nondisabled,* there is the risk of discrediting or devaluing the disabled experience—blanketing everyone under one group, instead of accounting for the individual. Therefore, in my attempt to avoid language that is unconsciously laden with particular meanings, I have chosen to rely on the terminology that is most frequently implemented in disability studies scholarship: *disabled* and *nondisabled*.

Before devoting the majority of this essay to examining important moments from the show itself, it is also necessary to briefly explain the two models of disability studies—medical and social—and how *Switched at Birth* conforms to the latter. The medical model, which was prevalent in the 1990s, aligns disability with abnormality—as not meeting standards either physically, mentally, or psychologically.[10] The belief that a disabled person is broken and needs to be fixed is reflected in the "overcoming narrative" which situates disability as something that must be overcome in order to achieve a happy and fulfilling life.[11] Conversely, the social model—which is more predominant among scholars today, though not often the media—depicts disability not as abnormal, but, as Ella R. Browning writes, "simply a different way of living in and experiencing the world, an expression of physical- or neuro-diversity that should be acknowledged and valued."[12] Additionally, the social model locates the source of disability not on the individual themselves but rather on the environment. The editors of *Framed: Interrogating Disability in the Media* explain: "In the social model the impairment is seen as much less important. Instead it is a disabling environment, the attitudes of others (not the disabled person), and institutional structures that are the problems requiring solution. Disability is thus not a fixed condition but a social construct and open to action and modification. One may have an impairment (or "condition") but in the right setting and with the right aids and attitudes one may not be disabled by it."[13] In other words, an unaccommodating environment, rather than a physical, mental, or emotional impairment, is the active disabler.

While in academia huge strides have been made in the area of disability studies, a relatively new yet ever-growing interdisciplinary field, the media has been slower to apply the more positive, social model position. Linton contends: "disabled people are rarely depicted on television, in films, or in fiction as being in control of their own lives—in charge or actively seeking out and obtaining what they want and need. More often, disabled people are depicted as pained by their fate, or, if happy, it is through personal triumph over their adversity. The adversity is not depicted as lack of opportunity, discrimination, institutionalization, and ostracism; it is the personal burden of their own body or means of functioning."[14] Although this is a troubling occurrence in and of itself, it is especially problematic because of audience: media communicates to the masses, while academia does not. If media continues to abide by the medical model, aligning disability with abnormality and regularly implementing well-worn tropes, then society will continue to Other those who are different. Shows like *Switched at Birth* and the evolution from

ABC Family to Freeform is about more than avoiding the confusing conflation of *family* with *safe*. It is about pushing the boundaries of how family and personhood are defined; it is about tackling difficult yet rewarding subject matter; and, it is about having a channel that reflects a diverse reality in a far more affecting way than its predecessors.

Complicating the Disability Narrative

While this essay focuses primarily on how deafness is constructed in the show, it is important to note how *Switched at Birth* positively depicts a range of other disabilities as well. Elisa (Zoey Deutch), the daughter of the candidate who is running against John for state senate, has bipolar disorder. Daphne's friend Sharee (Bianca Bethune) has an abusive and mentally ill mother, and one of Daphne's boyfriends, Campbell (RJ Mitte), has a spinal injury. Regina is an alcoholic who also ends up developing osteonecrosis in her hands, which prevents her from continuing to sign. Even Bay herself has to deal with a temporary disability: a severe hand injury that affects her career as an artist. Yet, perhaps the most provocative depiction of disability in the show is when Lily (Rachel Shenton), the girlfriend of Bay's brother Toby, becomes pregnant with a baby who tests positive for Down Syndrome. Lily's own brother has a genetic disorder, so she is fully aware of what a child with special needs requires. Although the subject of aborting babies with Down Syndrome has been widely debated in scholarship of the last decade, it is not often depicted in the media, and rarely with the maturity in which *Switched at Birth* manages.[15]

Spread over several episodes, Lily and Toby talk through the challenges of taking care of a baby with Down Syndrome. And, when having seemingly convinced each other that an abortion is the right option for them, Toby defends their decision to Daphne:

TOBY: … being deaf and having Down Syndrome are completely different, and I thought you didn't see yourself as disabled.
DAPHNE: You're right, they are not the same, but the world looks at us and thinks how sad, but it's not sad, it's just different … but I don't want you to think that different equals worse.
TOBY: I don't need a lecture on diversity right now…. I know that disabled people are valuable and they have the right to exist. That doesn't mean I'm the guy to raise one of them. This will ruin my life.
DAPHNE: It will change your life, but it doesn't mean it will ruin it. I think it will expand it.[16]

Daphne not only takes a bold pro-life stance but she also makes a strong case for the importance of diversity—not just in terms of disabled and nondisabled, but in the range of existing disabilities. Daphne's honesty, combined with Toby's visit to a daycare center with disabled children, causes him to rethink his decision, and he and Lily decide to keep the child. Nonetheless, *Switched at Birth* does not simply conclude the subject with a rosy outlook, but it continues to show Lily's doubts and fears and the need for a strong support group—thus complicating the narrative in significant ways.

Although Daphne is clearly a central character in the show, one of the many noteworthy aspects of *Switched at Birth* is that it effectively portrays a variety of deaf characters. For example, Natalie (Stephanie Nogueras), a girl who attends Carlton—a high school for deaf students that is attended by Daphne and eventually Bay as well—is a lesbian, and Travis (Ryan Lane), another deaf student, has a difficult home life. Emmett (Sean Berdy), Bay's on-again-off-again boyfriend, is often referred to as a deaf James Dean because of his leather jacket and his motorcycle. Emmett is also a skilled drummer who explains to the disbelieving that music is about feeling over hearing. Like the average high school kids that they are, the television series shows them throwing parties, being sexually active, pranking and bullying, attending school functions, and making mistakes and learning from them. The show also does a remarkable job of incorporating humor that serves to dispel "outsider" fears about the deaf community, such as when Daphne tells a prejudiced classmate: "I'd offer you one [a cookie] but you might catch being deaf."[17] This is remarkable progress from, when in 1988, John S. Schuchman wrote, "in Hollywood's view there is little or no humour in deafness."[18]

Not everyone's deafness is the same either. On a spring break trip to Mexico, Daphne learns that deaf individuals in other countries are even more disconnected from mainstream society than they are in America. Additionally, Noah (Max Lloyd-Jones), a character who features in the later seasons of the show, has Meniere's disease. Not only do viewers watch him struggling with the gradual loss of his hearing but also the difficulties of transitioning into the deaf community. Furthermore, many of the deaf actors and actresses in the show are deaf in real life to varying degrees. Katie Lynn Leclerc, the actress who plays Daphne, has Meniere's disease like the character Noah, and she has explained in interviews how actually having a hearing impairment has made her more comfortable in her relationship with the deaf community and in her role, of course, as Daphne.[19]

When Daphne first begins to interact with her biological family, one of the central lessons they have to learn is, as Regina tells them, "Deaf is not a bad word."[20] Initially John and Kathryn feel enormous guilt over Daphne's impairment, and they have to learn to accept that her deafness is simply part of her character and not pitiable or inspirational. This latter position traps the disabled character within the familiar "overcoming" narrative—a point which is addressed early in the show when a reporter is interviewing the family after the news of the switch. She says, "It's inspiring … to go deaf at that age and overcome so much,"[21] and later, in the news article itself, she writes of "how triumphantly Daphne is handling her handicap."[22] This disabled-as-martyr narrative is quickly dispelled, however, as the show works to complicate Daphne's character—not only in her personality and her experiences in the show, but also in terms of race. Before Daphne found out about the switch, she had spent her whole life believing that she was Latina. This tension is depicted most overtly when Daphne goes to interview for a college scholarship specifically for Latin Americans, as Regina insists that Daphne is Latina because of the culture in which she was raised. The very striking image of Daphne, with her pale skin and red hair, waiting with a group of five clearly Latina women raises interesting questions about ethnic identity by blood versus by environment.

Additionally, although in the beginning of the show Daphne appears to be unfailingly good, her character becomes increasingly complex. Just as Daphne's view of the Kennishes as the "perfect" family is eventually shattered, so is our view of the one-dimensional saintly or inspiring disabled person. One the one hand, Daphne is extremely intelligent and kind, a skilled cook, and a talented basketball player. On the other hand, she makes poor relationship choices, struggles to forgive and forget, and gets into trouble with the law. What is important here is how the development of Daphne's character allows viewers to see her foremost as a human being, and that is what positive disability representation consists of.

Deaf-Hearing Relationships

One of the most pervasive topics throughout *Switched at Birth* is deaf-hearing relationships—among friends and family, lovers, and strangers. This plays out in three major ways in the television series: first, in romantic relationships between hearing and deaf individuals; second, in the way deaf characters communicate with the hearing who do not sign; and third, through the cochlear implant debate.[23] Daphne regularly dates hearing people throughout the show—a choice which Emmett initially takes issue with, stating that a hearing girl "wouldn't understand my culture, my family, my perspective of the world."[24] Emmett's mother Melody, played by Marlee Matlin, the only deaf winner of an Academy Award for her role in *Children of a Lesser God* (1987), displays similar reservations about deaf-hearing relationships. However, what *Switched at Birth* proves, as both Emmett and his mother do end up having successful relationships with hearing individuals, is that in fact there are multiple ways of communicating and connecting, and that preconceived notions and black and white thinking do not hold up in a gray world.

Committed to complicating one-dimensional characterization, *Switched at Birth* offers a variety of different examples of how deaf people communicate, as not all are capable of reading lips or feel comfortable talking audibly. This is hardly a minor point as nearly three decades ago, Schuchman observed that "Hollywood cannot or will not deal with the issue of language in the deaf community."[25] While Daphne reads lips and speaks audibly, Emmett prefers to sign, or, when interacting with people who do not know sign language, he types messages on his phone. Many deaf individuals feel self-conscious about how their voice sounds since they are unable to hear it, and Daphne tells Bay about a time when she and Emmett rode on his motorcycle, screaming at the top of their lungs: "It was like for the first time, we didn't have to worry what our voices sounded like."[26] When Emmett is practicing his speech, we witness his embarrassment and his frustrations with trying to adapt to a society that favors the hearing.

Although the cochlear implant might seem like a natural "solution," this is a tool that has caused significant controversy in the deaf community—an issue that the show tackles both insightfully and sensitively. The cochlear implant is one of the first subjects that John brings up to Regina when he learns that Daphne is deaf. Regina replies, "You think she needs to be fixed, right? She's comfortable being deaf,"[27] to which John retorts, "Oh come on, no one wants to be deaf."[28] This is the heart of the debate: what does it mean to "fix" deafness? Cochlear implants are different than hearing aids because hearing aids only amplify sound, whereas cochlear implants enable people to hear by transferring sound signals to the brain—effectively avoiding the damaged part of the ear altogether. Therefore, this surgery raises a contentious question: When a deaf person receives a cochlear implant, are they no longer in the deaf community? Are they forever in a limbo between hearing and deaf? At various points in the show Daphne and Regina talk about why Regina chose not to get the cochlear implant for Daphne when she was a child—raising the additional question of what a parent of a deaf child should or should not do. The cochlear implant debate also manifests itself in Emmett's family when his dad, Cameron (Anthony Natale), decides to get the implant. At first Emmett does not handle the news well, and, when he tells Travis about it, Travis tells Emmett he himself was thinking about getting one:

> EMMETT: Hold up. You're deaf with a capital D.
> TRAVIS: I just think having a cochlear would open up more doors.
> EMMETT: Suddenly deaf isn't good enough for you?
> TRAVIS: Come on dude, that's so old school.[29]

Here, Emmett draws attention to the fact that many in the deaf community feel that there is a hierarchy of who is *Deafer* solely in terms of pride, not degree of deafness. It is worth noting, as well, that writing about deafness often formats the term *deaf* in various ways: sometimes the *d* is capitalized when referring to deaf culture and lowercased when referring to hearing loss; another option is to use *d/Deaf*.[30] Although this essay uses the lowercase *d* exclusively, it is important to note how these language choices speak to identity politics and deaf culture. Similarly, when Melody and Cameron (her ex-husband) talk about the implant, an important conversation occurs about self, as he wonders, "What if I lose my identity as a deaf person?"[31] The scene in which Cameron begins to first hear after receiving the implant is both affecting, as Cameron is able to audibly hear Emmett play drums for the first time, and humorous, as he tells Emmett that he is surprised that air conditioning, plastic bags, and drinking tea makes noise. What is powerful about this scene in particular is that both Cameron, with his cochlear implant, and Melody, without an implant, are standing together, smiling and enjoying Emmett's music—albeit in different ways.

Deaf Bodies in Public Spaces

In addition to how the show addresses actual matters that concern the deaf community—like deaf-hearing relationships and the cochlear implant debate—one of the major strengths of *Switched at Birth* is how it depicts Daphne navigating public spaces such as school and employment. Although Daphne is fortunate enough to attend Carlton, John and Kathryn immediately want her to go to Buckner Hall: a "regular school" with "regular kids."[32] While Daphne does remain at Carlton, she signs up to take a cooking class at Buckner—a move which highlights the difficulties and dangers of entering a space that is not willing to make accommodations. For example, her cooking teacher is not willing to slow down or face Daphne when giving instructions, and the traditional timer is of no use to Daphne, who ends up burning a recipe because she cannot hear when it is done. Similarly, when Daphne quits basketball at Carlton in order to play for Buckner, she comes to the realization that Buckner only wants her because she is deaf—a condition which enabled the school to receive a grant. By showing Daphne attending both Carlton and Buckner, the series illustrates the challenges (and unfortunate realities) of attending a school that is insensitive to those with special needs.

The Carlton and Buckner worlds collide, however, when Carlton implements a pilot program for hearing kids with deaf parents or siblings. Bay decides to leave Buckner and join Daphne at Carlton—a direction which serves to "normalize" or humanize the deaf characters even further, as the show does an incredible job of reversing what viewers have seen in the show thus far. With the pilot program, the deaf students become the ones angry at having to slow their classes down for the hearing kids who are having trouble keeping up. Bay gets called (signed) nasty words by her deaf classmates and feels hated simply because she is hearing. This reversal is important in that it forces viewers to confront their own behavior towards those with different abilities by seeing how unfair it is when the situation is inverted. When the only way to keep Carlton open is to accept more hearing students (which secures more funding), the deaf students protest—with the show making explicit comparisons to the Gallaudet University student protest in 1988. While the students do not win, and we see similar issues played out when Daphne transitions into college, the show manages to draw attention to a major problem in public education: the lack of funding, resources, and preparedness available for special needs students.

Additionally, the show does not shy away from representing the difficulties of securing employment as a disabled individual. Daphne has twelve interviews at various restaurants—each one dismissing her as soon as they find out she is deaf. Unable to obtain a job on her own, Daphne is forced to use Kathryn's connections. Getting the job proves to be the least of her difficulties, however, as Daphne knocks over dishes, messes up orders, and finally tells the chef that she needs to see his lips when he is speaking so that she can read them, to which he responds: "No I cannot.... I'm sure you grew up being told that you could do whatever you wanted ... but now I am going to tell you the truth. You can't do everything.... We weren't meant to do everything. That's why God created us different ... you're in my kitchen ... and now that's a problem. The law says I have to accommodate you to the best of my abilities so, if you want to stay, go wash dishes. You can't do much damage there."[33]

Eventually Daphne moves past the hurt and self-pity, and she decides to take matters into her own hands: creating a more accommodating space by putting up a mirror at work so that she can see behind herself. Nonetheless, when Daphne leaves the kitchen and takes her food truck out with Travis, other dangers arise. When they get robbed, the police do little beyond ignorantly stating that deaf individuals should not be out without supervision. Daphne's interaction with the police is mild, however, when compared with what Emmett endures in one of the most haunting scenes of *Switched at Birth*. When the cops show up at Emmett's house to ask him questions, they do not understand that he cannot hear them. They think that he is resisting, and they handcuff him—effectively silencing him. Afterwards Emmett signs to Daphne, "They had lights in my eyes. I couldn't see anything. I couldn't sign or speak. It was like I was drowning. I hated being deaf. And I hated myself for feeling that way."[34]

The lack of understanding and resources within civil service fields for the hearing impaired is also reflected in Daphne's decision to pursue a career in medicine, as there is an overwhelming need for deaf medical professionals. When several deaf students get into an accident, Daphne is forced to serve as a translator at the hospital. There are no doctors or nurses on call who know sign language, and the nurse explains that the hospital does not have a contract with a deaf interpreter on the weekends, to which Daphne exclaims, "so deaf people aren't supposed to get hurt on the weekend?"[35] Yet again, *Switched at Birth* is much more than a passive form of entertainment; it is actively drawing attention to the need for public spaces to accommodate diversity.

Switched at Birth: *Response and Influence*

In 1988, Schuchman confessed that "films continue to serve as a major source of public misinformation about deafness and deaf people. The deaf community awaits the next step in the industry's portrayal of deafness."[36] As I hope to have shown here, *Switched at Birth* has, for many, answered the long-awaited call for more accurate, complex depictions of deafness in mainstream media—especially in its aim at younger audiences. Schuchman was writing at a time when the disabled community first began to be recognized in significant ways by the public. In 1990, the Americans with Disabilities Act (ADA) became law, effectively prohibiting discrimination against those with disabilities, and scholars began to speak more openly about the representation of disability in mainstream media.

The year after the ADA was passed, Paul Hunt drew attention to the most common disabled stereotypes found in the media: "as pitiable or pathetic," "an object of curiosity or violence," "sinister or evil," "the super cripple," "as atmosphere," "laughable," "his/her own worst enemy," "as a burden," "as nonsexual," and as "being unable to participate in daily life."[37] Unfortunately, while some progress has been made with disability-positive shows like *Switched at Birth*, many of these stereotypes continue to be implemented in mainstream media. And, before concluding this essay, I aim to emphasize why this conversation about the need for more positive construction of ability diversity in television matters. Recognizing the role that media could play in decreasing fears and misunderstandings about those with disabilities, Colin Barnes in 1991 asserted, "Although there is some dispute about the level of influence the mass media has on our perceptions of the world, there are relatively few who would argue that it does not have any."[38] Worth reiterating, Barnes's statement was made in 1991—before streaming, DVRs, iPads and iPhones, and continual mass media input and output. *Switched at Birth* entertains this idea in interesting metatextual ways as well in their characterization of Emmett: a film enthusiast who goes on to study filmmaking in college. A decade after

Barnes's comment, Tom Shakespeare, with certainty, argues, "Such stereotypes reinforce negative attitudes towards disabled people, and ignorance about the nature of disability."[39] Accordingly, the question arises of where we are today in our understanding of how media affects thought, particularly when it comes to the representation of marginalized groups like the disabled.

In 2013, Lingling Zhang and Beth Haller conducted a study in which they asked a population of disabled individuals how they felt about American media representations of disability. Of their findings, Zhang and Haller note that the majority of responses suggested "that American media representations of people with disabilities were not realistic," and centered on depicting them as "sick," "relying on social or economic support," and "superhuman."[40] Furthermore, not only do Zhang and Haller observe how stereotypes affect the way nondisabled individuals view the disabled, but also how this influences the way the disabled view themselves—concluding "that mass media indeed can influence who we are and what we do."[41] Similarly, Katherine A. Foss, focusing specifically on representations of deafness in the media, asserts that even though hearing loss is one of the most common disabilities, "hearing loss receives little attention in popular and political discourse, except in its connection to aging."[42] Furthermore, Foss notes that if the individuals with hearing loss in the television programs are younger, most overcome their hearing loss—thus contributing to the overcoming narrative that has been discussed elsewhere in this paper.[43] Like others, Foss attests to the power of the media in informing our understandings of the world, and, in a separate but related article, Foss argues: "To change public perceptions, more counter messages need to appear in television, with storylines that do not specify 'deafness,' selecting d/Deaf actors of varying ethnicities and backgrounds, and conveying dialogue to a (hearing) audience, while preserving the voice of d/Deaf characters. Only through a diversity of Deaf characters and multiple storylines that present perspectives of the Deaf community can d/Deaf people be destigmatized."[44]

Echoing Foss, and reiterating much of what has already been stated here, *Switched at Birth*, in its diverse deaf characters, its portrayal of deaf culture, and its commitment to focusing on the *humanness* of the disabled, makes it a model show for combating stigmatization. If one needs further proof, in 2014, Seon-Kyoung An, Llewyn Elise Paine, and Jamie Nichole McNiel published an article titled "Prominent Messages in Television Drama *Switched at Birth* Promote Attitude Change Toward Deafness." After surveying a group of individuals both before and after watching *Switched at Birth*, and witnessing how the participants' perception of deafness was positively changed, they concluded: "This demonstrates that television media can play a role in forming people's attitudes on the subject of deafness and is consistent with both social cognitive theory and cultivation theory."[45]

While these theories confirm that outside forces, like the media, do indeed affect our worldviews, after watching *Switched at Birth* myself—as someone who is not hearing impaired—I attest that I did not need theory to validate my reaction. As a disability studies scholar, I am more sensitive than the average viewer to how disability is depicted in the media; however, *Switched at Birth* managed to surprise, unnerve, and alter my perspective in ways that I had not expected. Moments in the series in which the viewers saw the world from a deaf character's perspective—unable to hear background noise or voices—and the episode in season two that was presented entirely in ASL allowed me to tangibly experience a different way of seeing the world. For those who continue to dismiss the power of the media or disregard personal testimony, the evidence, as I have hoped to include to sufficient extent here, is available. Shows like *Switched at Birth* work to destigmatize disability, not only for the nondisabled, but in really important ways for the disabled community as well: access to positive representation encourages marginalized individuals to confidently claim their "non-normative" identities. As Rosemarie Garland-Thomson explains, "Becoming disabled demands learning how to live effectively as a person with disabilities ... [i]t means moving from isolation to community, from ignorance to knowledge about who we are, from exclusion to access, and from shame to pride."[46] Furthermore, the importance of this for young viewers in particular cannot be underestimated, as many disabled individuals, like Rosemarie Garland-Thomson herself, do not "come out" until much later in life. With the transition from ABC Family to Freeform, from long-held associations about what is "safe" and "normal" to deconstructing boundaries, I am optimistic that programming on this channel will inspire the media at large to accept its responsibility to further the acceptance of ability diversity.

NOTES

1. Lizzy Weiss, "This Is Not a Pipe," *Switched at Birth*, season 1, episode 1, ABC Family, June 6, 2011.
2. "Switched at Birth: Awards," *IMBD*, last modified 2017, http://www.imdb.com/title/tt1758772/awards/
3. "FAQs," *Freeform*, last modified 2016, http://freeform.go.com/faq.
4. Eric Deggans, "ABC Family Channel to Change Its Name to Freeform Network," *NPR*, January 12, 2016, http://www.npr.org/2016/01/12/462754255/abc-family-channel-to-change-its-name-to-freeform-network.
5. *Ibid.*
6. Nancy Mairs, "On Being a Cripple," in *The Norton Reader*, ed. Linda Peterson, John Brereton, Jospeh Bizup, Anne Fernald, and Melissa Goldthwaite (New York: Norton, 2012), 46.
7. *Ibid.*, 46–47.
8. Simi Linton, *Claiming Disability: Knowledge and Identity* (New York: New York University Press, 1998), 14–16.
9. Brenda Jo Brueggemann, Linda Feldmeier White, Patricia A. Dunn, Barbara A.

Heifferon, and Johnso Cheu, "Becoming Visible: Lessons in Disability," *College Composition and Communication* 52, no. 3 (2001): 369, accessed May 4, 2016, doi:10.2307.358624.

10. Ella R. Browning, "Disability Studies in the College Classroom," *Composition Studies* 42, no. 2 (2014): 99.

11. *Ibid.*

12. *Ibid.*

13. Ann Pointon and Chris Davies, introduction to *Framed: Interrogating Disability in the Media*, ed. Ann Pointon and Chris Davies (London: British Film Institute, 1997), 2.

14. Simi Linton, *Claiming Disability: Knowledge and Identity*, 25.

15. Chris Kaposy, "A Disability Critique of the New Prenatal Test for Down Syndrome," *Kennedy Institute of Ethics Journal* 23, no. 4 (2013): 299–324.

16. Bekah Brunstetter, "How Does a Girl Like You Get to Be a Girl Like You," *Switched at Birth*, season 4, episode 12, ABC Family, August 31, 2015.

17. Henry Robles, "The Persistence of Memory," *Switched at Birth*, season 1, episode 6, ABC Family, July 11, 2011.

18. John S. Schuchman, "Deafness and the Film Entertainment Industry," in *Hollywood Speaks: Deafness and the Film Entertainment Industry* (Urbana: University of Illinois Press, 1988), 44.

19. Robin Hilmantel, "Why Katie Leclerc Says Her Intermittent Hearing Loss Is a 'Huge Blessing,'" *Women's Health*, February 3, 2014, http://www.womenshealthmag.com/health/menieres-disease.

20. Lizzy Weiss, "This Is Not a Pipe."

21. Chad Fiveash and James Stoteraux, "Dance Amongst Daggers," *Switched at Birth*, season 1, episode 4, ABC Family, June 27, 2011.

22. Joy Gregory, "Dogs Playing Poker," *Switched at Birth*, season 1, episode 5, ABC Family, July 4, 2011.

23. Lauren Pass and Abraham D. Graber, "Informed Consent, Deaf Culture, and Cochlear Implants," *The Journal of Clinical Ethics* 26, no. 3 (2015): 219–230.

24. Becky Hartman Edwards, "Portrait of My Father," *Switched at Birth*, season 1, episode 3, ABC Family, June 20, 2011.

25. John S. Schuchman, "Deafness and the Film Entertainment Industry," 45.

26. Joy Gregory, "Les Soeurs d'Estrées," *Switched at Birth*, season 1, episode 14, ABC Family, January 24, 2012.

27. Lizzy Weiss, "This Is Not a Pipe," *Switched at Birth*.

28. *Ibid.*

29. Lizzy Weiss and Michael Ross, "He Did What He Wanted," *Switched at Birth*, season 2, episode 14, ABC Family, July 1, 2013.

30. Jamie Berke, "Deaf Culture—Big D Small D," *Very Well*, June 28, 2016, https://www.verywell.com/deaf-culture-big-d-small-d-1046233.

31. Terrence Coli & Becky Hartman Edwards, "Prudence, Avarice, Lust, Justice, Anger," *Switched at Birth*, season 2, episode 17, ABC Family, July 22, 2013.

32. Lizzy Weiss, "This Is Not a Pipe," *Switched at Birth*.

33. Lizzy Weiss, "This Is the Color of My Dreams," *Switched at Birth*, season 1, episode 23, ABC Family, September 3, 2012.

34. Becky Hartman Edwards, "Self-Portrait with Bandaged Ears," *Switched at Birth*, season 1, episode 13, ABC Family, January 17, 2012.

35. Lizzy Weiss and Bekah Brunstetter, "There Is My Heart," *Switched at Birth*, season 4, episode 10, ABC Family March 10, 2015.

36. John S. Schuchman, "Deafness and the Film Entertainment Industry," 48.

37. Lucy Wood, "Media Representation of Disabled People," *Disability Planet*, 2012, http://www.disabilityplanet.co.uk/critical-analysis.html.

38. Colin Barnes, "Discrimination: Disabled People and the Media," originally published in *Contact* 70 (1991): 45–48. Quote is taken from page 2 of the online PDF document, accessed October 11, 2016, http://www.leeds.ac.uk/disability-studies/archiveuk/archframe.htm.

39. Tom Shakespeare, "Art and Lies? Representations of Disability on Film," in *Disability*

Discourse, ed. Mairian Corker and Sally French (Buckingham: Open University Press, 1999), 166.

40. Lingling Zhang and Beth Haller, "Consuming Image: How Mass Media Impact the Identity of People with Disabilities," *Communication Quarterly* 61, no. 3 (2013), 326.

41. Lingling Zhang and Beth Haller, "Consuming Image: How Mass Media Impact the Identity of People with Disabilities," 330.

42. Katherine A. Foss, Abstract to "(De)stigmatizing the Silent Epidemic: Representations of Hearing Loss in Entertainment Television," *Health Communication* 29 (2014): 888.

43. *Ibid.*, 893.

44. Katherine A. Foss, "Constructing Hearing Loss or 'Deaf Gain?' Voice, Agency, and Identity in Television's Representations of d/Deafness," *Critical Studies in Media Communication* 31, no. 5 (2014): 445.

45. Seon-Kyoung An, Llewyn Elise Paine, Jamie Nichole McNiel, Amy Rask, Jourdan Taylor Holder, and Duane Varan, "Prominent Messages in Television Drama Switched at Birth Promote Attitude Change Toward Deafness," *Mass Communication and Society* 17 (2014): 210.

46. Rosemarie Garland-Thomson, "Becoming Disabled," *The New York Times*, August 19, 2016, http://www.nytimes.com/2016/08/21/opinion/sunday/becoming-disabled.html.

Bibliography

Allen, Robert C., and Annette Hill, eds. *The Television Studies Reader*. New York: Routledge, 2003.

An, Seon-Kyoung, Llewyn Elise Paine, Jamie Nichole McNiel, Amy Rask, Jourdan Taylor Holder, and Duane Varan. "Prominent Messages in Television Drama Switched at Birth Promote Attitude Change Toward Deafness." *Mass Communication and Society* 17, no. 2 (2014): 195–216.

Banet-Weiser, Sarah, Cynthia Chris, and Anthony Freitas, eds. *Cable Visions: Television Beyond Broadcasting*. New York: New York University Press, 2007.

Benedict, Helen. *Virgin or Vamp: How the Press Covers Sex Crimes*. New York: Oxford University Press, 1992.

Bingham, Patrick. "*Pretty Little Liars*: Teen Mystery or Revealing Drama?" *Networking Knowledge* 6, no. 4 (2014): 95–106.

Bradley, Laura. "ABC Family Is Smart to Change Its Name. Maybe Other Networks Should Follow Suit, Too." *Slate Magazine*, October 6, 2015, http://www.slate.com/blogs/brow beat/2015/10/06/abc_family_s_name_change_to_freeform_is_smart.html.

Brown, Shane. "Pushing Boundaries? Challenging Traditional Gender Roles in *Kyle XY*." *Science Fiction Film and Television* 8, no. 1 (2015): 91–100.

Brunsdon, Charlotte, Julie D'Acci, and Lynn Spiegel, eds. *Feminist Television Criticism: A Reader*. Oxford: Oxford University Press, 1997.

Churchill, Gilbert A., and George P. Moschis. "Television and Interpersonal Influences on Adolescent Consumer Learning." *Journal of Consumer Research* 6, no. 1 (1979): 23–35.

Cuklanz, Lisa M. *Rape on Prime Time: Television, Masculinity, and Sexual Violence*. Philadelphia: University of Pennsylvania Press, 2000.

D'Acci, Julie. *Defining Women: Television and the Case of Cagney and Lacey*. Chapel Hill: University of North Carolina Press, 1994.

_____, ed. "Lifetime: A Cable Network 'for Women.'" Special issue, *Camera Obscura: A Journal of Feminism, Culture, and Media Studies* 33–34 (1994–1995).

Davis, Glyn, and Gary Needham. *Queer TV: Theories, Histories, Politics*. New York: Routledge, 2009.

Davis, Glyn, and Kay Dickinson, eds. *Teen TV: Genre, Consumption and Identity*. London: British Film Institute, 2004.

Douglas, Susan J. *Enlightened Sexism: The Seductive Message That Feminism's Work is Done*. New York: Times Books, 2010.

_____. *Where the Girls Are: Growing Up Female with the Mass Media*. New York: Three Rivers Press, 1994.

Dow, Bonnie J. *Prime Time Feminism: Television, Media Culture, and the Women's Movement Since 1970*. Philadelphia: University of Pennsylvania Press, 1996.

Drucker, Susan J., and Gary Gumpert, eds. *Voices in the Street: Explorations in Gender, Media, and Public Space*. Cresskill, NJ: Hampton Press, 1996.

Ellcessor, Elizabeth. *Restricted Access: Media, Disability, and the Politics of Participation*. New York: New York University Press, 2016.

Feasey, Rebecca. "Charmed: Why Teen Television Appeals to Women." *Journal of Popular Film and Television* 34, no. 1 (2006): 2–9.

Foss, Katherine A. "Constructing Hearing Loss or 'Deaf Gain?' Voice, Agency, and Identity in Television's Representations of d/Deafness." *Critical Studies in Media Communication* 31, no. 5 (2014): 426–447.

_____. "(De)stigmatizing the Silent Epidemic: Representations of Hearing Loss in Entertainment Television." *Health Communication* 29 (2014): 888–900.

Gauntlett, David. *Media, Gender and Identity: An Introduction*. New York: Routledge, 2008.

Genz, Stephanie. *Postfemininities in Popular Culture*. Hampshire: Palgrave Macmillan, 2009.

Genz, Stephanie, and Benjamin A. Brabon. *Postfeminism: Cultural Texts and Theories*. Edinburgh: Edinburgh University Press, 2009.

Gill, Rosalind. *Gender and the Media*. Cambridge: Polity, 2007.

Gillan, Jennifer. *Television Brandcasting: The Return of the Content Promotion Hybrid*. New York: Routledge, 2015.

Gillig, Traci K., and Sheila T. Murphy. "Fostering Support for LGBTQ Youth? The Effects of a Gay Adolescent Media Portrayal on Young Viewers." *International Journal of Communication* 10 (2016): 3828–3850.

Greene, Doyle. *Teens, TV and Tunes: The Manufacturing of American Adolescent Culture*. Jefferson, NC: McFarland, 2012.

Hargreaves, Jennifer. *Sporting Females: Critical Issues in the History and Sociology of Women's Sports*. New York: Routledge, 1994.

Howe, Neil, and William Strauss. *Millennials Rising: The Next Great Generation*. New York: Vintage Original, 2000.

Hundley, Heather. "The Evolution of Gendercasting: The Lifetime Television Network—'Television for Women.'" *Journal of Popular Film and Television* 29, no. 4 (2002): 174–181.

James, Carrie. *Disconnected: Youth, New Media, and the Ethics Gap*. Cambridge: MIT Press, 2014.

Jenkins, Henry. *Textual Poachers: Television Fans & Participatory Culture*. New York: Routledge, 1992.

Johnson, Merri Lisa, ed. *Third Wave Feminism and Television: Jane Puts It in a Box*. London: I.B. Tauris, 2007.

Kaveney, Roz. *Teen Dreams: Reading Teen Film from Heathers to Veronica Mars*. London: I.B. Tauris, 2006.

Kearney, Mary Celeste, ed. *The Gender and Media Reader*. New York: Routledge, 2011.

Keller, James R. *Queer (Un)Friendly Film and Television*. Jefferson, NC: McFarland, 2002.

Kelly, Maura. "Virginity Loss Narratives in 'Teen Drama' Television Programs." *Journal of Sex Research* 47, no. 5 (2010): 479–489.

Kenny, Lorraine Delia. *Daughters of Suburbia: Growing Up White, Middle Class and Female*. New Brunswick: Rutgers University Press, 2000.

Kielwasser, Alfred P., and Michelle A. Wolf. "Mainstream Television, Adolescent Homosexuality, and Significant Silence." *Critical Studies in Mass Communication* 9, no. 4 (1992): 350–373.

Kohnen, Melanie E. S. "Cultural Diversity as Brand Management in Cable Television." *Media Industries Journal* 2, no. 2 (2015): 88–103.

Levine, Elana. *Cupcakes, Pinterest, and Ladyporn: Feminized Popular Culture in the Early Twenty-First Century*. Urbana: University of Illinois Press, 2015.

Levy, Ariel. *Female Chauvinist Pigs: Women and the Rise of Raunch Culture*. New York: Free Press, 2005.

Lotz, Amanda. *Beyond Prime Time: Television Programming in the Post-Network Era*. New York: Routledge, 2009.

_____. *Cable Guys: Television and Masculinities in the 21st Century*. New York: New York University Press, 2014.

_____. *Redesigning Women: Television after the Network Era*. Urbana: University of Illinois Press, 2006.

_____. *The Television Will Be Revolutionized.* 2d ed. New York: New York University Press, 2014.

Marwick, Alice E. *Status Update: Celebrity, Publicity, and Branding in the Social Media Age.* New Haven: Yale University Press, 2013.

McRobbie, Angela. "Young Women and Consumer Culture." *Cultural Studies* 22, no. 5 (2008): 531–550.

Milestone, Katie, and Anneke Meyer. *Gender and Popular Culture.* Cambridge: Polity Press, 2012.

Mitchell, Jennifer. "The Best Lesbian Show Ever!: The Contemporary Evolution of Teen Coming-Out Narratives." *Journal of Lesbian Studies* 19, no. 4 (2015): 454–469.

_____. *Genre and Television: From Cop Shows to Cartoons in American Culture.* New York: Routledge, 2004.

Mittell, Jason. *Complex TV: The Poetics of Contemporary Television Storytelling.* New York: New York University Press, 2015.

Mullen, Megan. "The Fall and Rise of Cable Narrowcasting." *Convergence: The International Journal of Research into New Media Technologies* 8, no. 1 (2002): 62–83.

_____. *The Rise of Cable Programming in the United States: Revolution or Evolution?* Austin: University of Texas Press, 2003.

Mulvey, Laura. "Visual Pleasure and Narrative Cinema." *Screen* 16, no. 3 (1975): 6–18.

Munford, Rebecca, and Melanie Waters. *Feminism and Popular Culture: Investigating the Postfeminist Mystique.* London and New York: I.B. Tauris, 2014.

Murphy, Caryn. "Secrets and Lies: Gender and Generation in the ABC Family Brand." In *The Millennials on Film and Television: Essays on the Politics of Popular Culture,* edited by Betty Kaklamanidou and Margaret Tally, 15–30. Jefferson, NC: McFarland, 2014.

Murray, Susan, and Laurie Ouellette, eds. *Reality TV: Remaking Television Culture.* New York: New York University Press, 2004.

Newcomb, Horace, ed. *Television: The Critical View.* 6th ed. Oxford: Oxford University Press, 2000.

Peters, Wendy. "Bullies and Blackmail: Finding Homophobia in the Closet on Teen TV." *Sexuality and Culture* 20 (2016): 486–503.

Polletta, Francesca, and Christine Tomlinson. "Date Rape After the Afterschool Special: Narrative Trends in the Televised Depiction of Social Problems." *Sociological Forum* 29, no. 3 (September 2014): 527–548.

Pozner, Jennifer L. *Reality Bites Back: The Troubling Truth About Guilty Pleasure TV.* Berkeley, CA: Seal Press, 2010.

Press, Andrea L. *Women Watching Television: Gender, Class, and Generation in the American Television Experience.* Philadelphia: University of Pennsylvania Press, 1991.

Projansky, Sarah. *Spectacular Girls: Media Fascination and Celebrity Culture.* New York: New York University Press, 2014.

_____. *Watching Rape: Film and Television in Postfeminist Culture.* New York: New York University Press, 2001.

Proulx, Mike, and Stacey Shepatin. *Social TV: How Marketers Can Reach and Engage Audiences by Connecting Television to the Web, Social Media, and Mobile.* Hoboken, NJ: John Wiley and Sons, 2012.

Radner, Hilary. *Neo-Feminist Cinema.* New York: Routledge, 2011.

Ross, Sharon Marie, and Louisa Ellen Stein, eds. *Teen Television: Essays on Programming and Fandom.* Jefferson, NC: McFarland, 2008.

Rowe, Kathleen. *The Unruly Woman: Gender and the Genres of Laughter.* Austin: University of Texas Press, 1995.

Sepinwall, Alan. *The Revolution was Televised: The Cops, Crooks, Slingers, and Slayers Who Changed TV Drama Forever.* New York: Touchstone, 2012.

Shary, Timothy. *Generation Multiplex: The Image of Youth in Contemporary American Cinema.* Austin: University of Texas Press, 2002.

Shary, Timothy, and Alexandra Seibel, eds. *Youth Culture in Global Cinema.* Austin: University of Texas Press, 2007.

Slade, Alison F., Amber J. Narro, and Dedria Givens-Carroll, eds. *Television, Social Media, and Fan Culture.* Maryland: Lexington Books, 2015.

Spiegel, Lynn. *Make Room for TV: Television and the Family Ideal in Postwar America.* Chicago: University of Chicago Press, 1992.

Squires, Catherine R. *The Post-Racial Mystique: Media & Race in the Twenty-First Century.* New York: New York University Press, 2014.

Stein, Louisa Ellen. *Millennial Fandom: Television Audiences in the Transmedia Age.* Iowa City: University of Iowa Press, 2015.

_____. "'Word of Mouth on Steroids': Hailing the Millennial Media Fan." In *Flow TV: Television in the Age of Media Convergence*, edited by Michael Kackman, Marnie Binfield, Matthew Thomas Payne, Allison Perlman, and Bryan Sebok, 128–143. New York: Routledge, 2010.

Tasker, Yvonne, and Diane Negra, eds. *Interrogating Postfeminism: Gender and Politics of Popular Culture.* Durham: Duke University Press, 2007.

Thompson-Spires, Nafissa D. "Tolerated, but Not Preferred: Troubling the Unconscious of Televisual Multiculturalism." *American Review of Canadian Studies* 41, no. 3 (September 2011): 293–307.

Tukachinsky, Riva, Dana Mastro, and Moran Yarchi. "Documenting Portrayals of Race/Ethnicity on Primetime Television Over a 20-Year Span and Their Association with National-Level Racial/Ethnic." *Journal of Social Issues* 71, no. 1 (2015): 17–38.

Zeisler, Andi. *Feminism and Pop Culture.* Berkeley: Seal Press, 2008.

Zhang, Lingling, and Beth Haller. "Consuming Image: How Mass Media Impact the Identity of People with Disabilities." *Communication Quarterly* 61, no. 3 (2013): 319–334.

About the Contributors

Cara **Dickason** is a doctoral candidate in screen cultures at Northwestern University. Her research explores the intersection of surveillance, spectatorship, and gender in contemporary girls' and women's television. She previously taught English and composition at Georgetown University, Trinity Washington University, and Prince George's Community College.

Anne **Dotter** is the associate director of the Honors Program at the University of Kansas. Her research focuses on critical race theory applications on analyses of gender representations, cultural translation and issues tied to Honors education in the United States, including questions of diversity and inclusion.

Anelise **Farris** is a doctoral candidate in English at Idaho State University specializing in disability studies, literature of the fantastic, and folklore. Her research focuses on disability in virtual spaces. Her past publications have considered various ways in which folklore, pop culture, and disability studies intersect.

Jessica **Ford** is an early career researcher at UNSW Sydney. She is a co-founder of the Sydney Screen Studies Network, a community of scholars and researchers. Her work examines women and feminism on screen and she has published on various female-centric U.S. television series, including *Buffy the Vampire Slayer* and *Girls*.

Malynnda A. **Johnson** is an assistant professor of communication at Indiana State University. With a background in health communication, she specializes in media representations of health and the critical examinations of entertainment education. She also studies prevention and intervention programs for community organizations.

Erica **Lange** is a doctoral student in rhetoric and composition and instructor of composition studies in the English department at Ohio University. She has pedagogical concentrations in food culture and cultural competencies. Her research includes composition pedagogy, rhetorical theory, feminism, and bodily rhetorics.

Joe **Lipsett** is an educational developer in faculty development at OCAD University in Toronto. He has taught film courses on slashers, puzzles and dystopias as well as cult films and the impact of technology on society. His research focuses on horror films and teen television.

Andi **McClanahan** is a professor of communication and the coordinator of the Women & Gender Studies Program at East Stroudsburg University of Pennsylvania. She specializes in rhetorical and communication theory as well as critical media and gender studies.

Nikki Jo **McCrady** is a graduate of the University of Southern Indiana, where she majored in communication studies. She is the program communications and events specialist for Big Brothers Big Sisters of Central Indiana.

Mel **Medeiros** is a doctoral candidate at Michigan State University in the media and information program, where she studies political communication and new technology, specifically focusing on political polarization and information flows. Her past research has focused on portrayals of foster care.

Emily L. **Newman** is an associate professor of art history at Texas A&M University–Commerce, specializing in contemporary art, gender studies, and popular culture. She is the coeditor, with Emily Witsell, of *The Lifetime Network* (2016) and the author of many articles/reviews in art history and popular culture journals.

Donica **O'Malley** is a doctoral candidate in the Department of Communication at the University of Pittsburgh, focusing on studies of media and popular culture. Her dissertation traces the "ginger phenomenon" in social media and analyzes the perception of red hair and pale skin as a social difference.

Patrice A. **Oppliger** is an assistant professor of mass communication at Boston University. Her research interests include media effects, particularly for adolescents, and humor studies. She is the author of *Wrestling and Hypermasculinity* (2004); *Girls Gone Skank* (2008); and *Bullies and Mean Girls in Popular Culture* (2013).

Sharon L. **Pajka** is a professor in the Department of English at Gallaudet University where she teaching courses in adolescent literature and literary studies. She publishes the *Deaf Characters in Adolescent Literature* blog, which highlights publications and author interviews.

Madeline **Rislow** is an assistant professor and director of art history at Missouri Western State University. While she is a specialist in fifteenth-century Genoese sculpture, her research and teaching interests include everything from prehistoric art to contemporary art and popular culture.

Stephen P. **Smyth** is an independent scholar. His work focuses on post-war architecture, mass media, and American popular culture. He has presented at academic conferences around the U.S. and chairs the Architecture and Built Environment Area for the Mid-Atlantic Popular and American Culture Association.

Kathleen M. **Turner** spent many years as faculty in academia and is now a teacher at Jefferson City High School. She is also the executive secretary of the Midwest Popular Culture Association/American Culture Association.

Emily **Witsell** is a research librarian and coordinator of reference and instruction at Wofford College. She is the coeditor, with Emily L. Newman, of *The Lifetime*

Network (2016). Her past research has focused on the cultural constructions of illness.

Stephanie L. **Young** is an associate professor of communication studies at the University of Southern Indiana, specializing in rhetorical criticism, autoethnography, and issues of race, gender, and sexuality in popular culture. She has published in a number of scholarly journals.

Index